AF564541

HORTICULTURE

NIPA GENX ELECTRONIC RESOURCES & SOLUTIONS P. LTD.
New Delhi-110 034

Dr. Pranava Pandey is Assistant Professor-cum-Junior Scientist (Horticulture-Fruit Science) at the Department of Horticulture, Veer Kunwar Singh College of Agriculture, Dumraon, Bihar Agricultural University, Sabour, Bhagalpur, Bihar. He has done B.Sc. (Agri.) from NDUA&T, Ayodhya and M.Sc. (Ag.) Horticulture from Banaras Hindu University (BHU), Varanasi and Doctorate (Ph.D.) in Horticulture (Pomology) from Indian Agricultural Research Institute (IARI), New Delhi. He was awarded ICAR-SRF (PGS) in Fruit Science. He has more than 7 years of experience in teaching, research, extension and training. He has published several research papers in reputed peer reviewed national and international journals, book chapters, technical bulletins, extension folder and popular articles. His research work is oriented towards crop production and improvement of fruit crops.

Dr. Avnish Kumar Pandey is currently working as Assistant Professor, Department of Fruit Science, ASPEE College of Horticulture and Forestry, Navsari Agricultural University, Navsari, Gujarat. He has completed his B.Sc. (Hons.) Agriculture from CSAUA&T, Kanpur, (U.P.), M.Sc. Fruit Science (Hort.) from Dr. YSPUH&F, Solan, (H.P.) and received ICAR-JRF award and further completed Ph.D. in Fruit Science (Hort.) from NDUA&T, Ayodhya, (U.P.). He has more than 5 years of experience in teaching, research and extension. Dr. Pandey has supervised eight P.G. students and handled several research projects. He has been awarded Young Scientist Award (2017). He has published more than 35 research papers national and international reputed journal, 3 review papers, 15 book chapters, 7 popular articles and 1 book. He is member of several national and international scientific societies.

Dr. Sanjeev Kumar, is Assistant Professor in the Department of Vegetable Science at ASPEE College of Horticulture & Forestry, Navsari Agricultural University, Navsari, Gujarat and has professional experience of more than 17 years. He has worked in different capacity in various private and public organizations before joining to the present post. He completed graduation (Agri.), Master's (2001) and doctorate degree (2006) in the discipline of Vegetable Science with "Certificate of Honour" in Master's programme from CSK Himachal Pradesh Agricultural University, Palampur 176062 (HP), India and has guided 9 M.Sc. students in various aspects of protected cultivation and currently guiding 1 M.Sc. and 2 Ph.D. students. He is handling one In-house project on "Research in Vegetable Crops under Protected Conditions Phase-II" Co-PI and also PI of a project "Vegetable Grafting to Mitigate Biotic and Abiotic Stresses in Vegetable Crops" under RKVY. He has published 47 research papers, more than 45 popular articles in vernacular languages and contributed to 11 recommendations for farming community. He has delivered more than 35 lectures as resource person in seminar, training programmes *etc*. He has undergone more than 15 competence building programmes. He has contributed to formation of Vision-2050 of the College, 3 farmers-oriented booklets and written 3 books. He has also been recognized by ASM Foundation, New Delhi for Keynote presentation in a Global Conference during 2014. Dr Sanjeev Kumar is recipient of Young Scientist Award, 2016 and Best Scientist National Award, 2018. He is affiliated to 6 professional societies and 2 academic bodies, and is patron member of 'The Horticultural Society of Gujarat'. He organized a national seminar on "Technologies and Sustainability of Protected Cultivation for Hi-Valued Vegetable Crops" during February 01-03, 2018. He was also a 'Certified Trainer' of ASCI for protected cultivation and has successfully conducted skill development training as Training Coordinator during 2018.

HORTICULTURE
Basic Principles and Recent Approaches

Pranava Pandey
Assistant Professor-cum-Junior Scientist
Department of Horticulture
Veer Kunwar Singh College of Agriculture, Dumraon
Bihar Agricultural University
Sabour, Bhagalpur, Bihar

A.K. Pandey
Assistant Professor
Department of Fruit Science
ASPEE College of Horticulture and Forestry
Navsari Agricultural University
Navsari, Gujarat

Sanjeev Kumar
Assistant Professor
Department of Vegetable Science
ASPEE College of Horticulture & Forestry
Navsari Agricultural University
Navsari, Gujarat

NIPA GENX ELECTRONIC RESOURCES & SOLUTIONS P. LTD.
New Delhi-110 034

NIPA GENX ELECTRONIC
RESOURCES & SOLUTIONS P. LTD.

101,103, Vikas Surya Plaza, CU Block
L.S.C.Market, Pitam Pura, New Delhi-110 034
Ph : +91 11 27341616, 27341717, 27341718
E-mail:newindiapublishingagency@gmail.com
www: www.nipabooks.com

For customer assistance, please contact
Phone: + 91-11-27 34 17 17 Fax: + 91-11- 27 34 16 16
E-Mail: feedbacks@nipabooks.com

ISBN: 978-93-91383-96-1

Composed and Designed by NIPA.

Preface

Horticultural crops play an important role in nutritional security, economic viability and are very well adopted into the predominant intensive cropping systems prevailing in different parts of our country. India continues to be the second largest producer of fruits and vegetables across the globe, however, productivity of most of the horticultural commodities is very marginal and much lower in comparison to global trends and is still a challenge and a matter of great concern. The export earnings from fresh and processed fruits and vegetables, floriculture products and horticultural seed and planting material have achieved greater heights during last decade.

Of late, there has been a shift in the consumption pattern of fruits and vegetables and accordingly the supply side has to gear up to meet the demand requirements in terms of quantity, quality and seasonality. Horticultural crops are *per se* rich in essential nutrients, vitamins, minerals, dietary fiber *etc.* and have the potential to bring down the magnitude of malnourished population and improve the health of children in the country. With the projected population of 1.62 billion by 2050 and considering the standard dietary requirement of 400 g per capita, the domestic requirement of fruits and vegetables would be 400 million tonnes. However, taking into account post-harvest losses of fruits and vegetables to the tune of 20%, it is estimated that the production should at least be 480 million tonnes. Considering the growing demand for processed products and potential exports, the production target should be around 500 to 600 million tonnes. The increase in demand will be faced with challenges such as dwindling natural resources particularly land, irrigation water and conventional energy sources. Horticultural crops with their high productivity will play a crucial role in boosting food production and thereby the availability.

A significant change in climate on a global scale has greater impact on current agricultural practices and consequently food supply is being affected all around. Climate change *per se* is not necessarily harmful; the problems arise from extreme events that are difficult to predict. So, identification of climate-resilient horticultural crops should also be dealt positively in anticipation to such change.

Thus, the efforts have been made so to compile the book entitled "Horticulture: Basic Principles and Recent Approaches" for students, teachers, research scholars as well as budding horticulturists. This book covers most of the areas of horticulture including from traditional to recent advances. The emphasis has not only laid on definitions and technical terms, but also on the concepts and their applications. Special care has been taken so that the rigour of science is not lost while simplifying the language. We hope and trust that this book will help the students to understand basic and advance techniques and their utility in a very simple way. An attempt has been made to design chapters of the book based on the current curricula and syllabi of ICAR.

We are thankful to our respective home institutes; VKS College of Agriculture, Dumraon and Bihar Agricultural University, Sabour, Bhagalpur, Bihar and Navsari Agricultural

University, Navsari for invariable motivation in the preparation of this text. The constant encouragement and constructive criticism have motivated us to bring this manuscript to the present form. We are very grateful for a number of friends and colleagues in encouraging us to start the work, persevere with it and finally to publish the book. Devotion and dedication of esteemed authors of different chapters of the book is highly acknowledged for preparing, reviewing, refining and finalization of the manuscript in a set pattern and helping to bring it in the presentable form. We are highly thankful to the publisher for accepting our proposal and publishing the book without any complexities and delay.

Last but not the least, we owe the obvious support and sacrifice of our families. It would be unrealistic to assume that the book is free from errors and the suggestions will be highly appreciated for making improvement in this manuscript.

Editors

Contents

List of Contributors

Alka Singh, *Department of Floriculture and Landscape Architecture, ASPEE College of Horticulture and Forestry, Navsari Agricultural University, Navsari, Gujarat*

Amit Kumar, *Division of Fruit Science, Faculty of Horticulture SKUAST-Kashmir, Shalimar Campus, Srinagar, Jammu and Kashmir*

Ankit Pandey, *College of Horticulture, Rajmata Vijayaraje Scindia Krishi Vishwavidyalaya Campus Mandsaur, Madhya Pradesh*

A.K. Pandey, *ASPEE College of Horticulture and Forestry, Navsari Agricultural University, Navsari, Gujarat*

Dhara J. Barot, *College of Horticulture, S.D. Agricultural University, Jagudan, Gujarat*

Dushyant D. Champaneri, *Department of Vegetable Science, ASPEE College of Horticulture and Forestry, Navsari Agricultural University, Navsari, Gujarat*

H.V. Pandya, *Department of Plant Protection, ASPEE College of Horticulture and Forestry, Navsari Agricultural University, Navsari, Gujarat*

Hemant Sharma, *Department of Plant Protection, ASPEE College of Horticulture and Forestry, Navsari Agricultural University, Navsari, Gujarat*

Hetal Rathod, *ASPEE College of Horticulture and Forestry, Navsari Agricultural University, Navsari, Gujarat*

Jagdeesh Prasad Rathore, *Division of Fruit Science, Faculty of Horticulture, SKUAST-Kashmir, Shalimar Campus, Srinagar, Jammu and Kashmir*

M.L. Jat, *College of Horticulture, S.D. Agricultural University, Jagudan, Gujarat*

Madineni Tejaswini, *Department of Vegetable Science, ASPEE College of Horticulture & Forestry Navsari Agricultural University, Navsari ,Gujarat*

Mital Dudhat, *Department of Vegetable Science, ASPEE College of Horticulture & Forestry Navsari Agricultural University, Navsari, Gujarat*

Mukesh Kumar, *College of Horticulture, S.D. Agricultural University, Jagudan, Gujarat.*

P.R. Patel, *Department of Plant Protection, ASPEE College of Horticulture and Forestry, Navsari Agricultural University, Navsari, Gujarat*

Pranava Pandey, *Department of Horticulture, Veer Kunwar Singh College of Agriculture, Dumraon, Bihar Agricultural University, Sabour, Bhagalpur, Bihar*

R.K. Jat, *College of Horticulture, S.D. Agricultural University, Jagudan, Gujarat*

S.K. Acharya, *College of Horticulture, S.D. Agricultural University, Jagudan, Gujarat*

S.N. Saravaiya, *Department of Vegetable Science, ASPEE College of Horticulture & Forestry Navsari Agricultural University, Navsari, Gujarat*

Sanjeev Kumar, *Department of Vegetable Science, ASPEE College of Horticulture & Forestry Navsari Agricultural University, Navsari, Gujarat*

Shailendra K. Dwivedi, *College of Horticulture, Rajmata Vijayaraje Scindia Krishi Vishwavidyalaya Campus, Mandsaur, Madhya Pradesh*

Snehal M. Patel, *Department of Plant Protection, ASPEE College of Horticulture and Forestry, Navsari Agricultural University, Navsari, Gujarat*

T.R. Ahlawat, *Department of Fruit Science, ASPEE College of Horticulture and Forestry, Navsari Agricultural University, Navsari, Gujarat*

V.P. Prajapati, *Department of Plant Protection, ASPEE College of Horticulture and Forestry, Navsari Agricultural University, Navsari, Gujarat*

Vivek Tiwari, *Defence Institute of High Altitude Research (DIHAR), DRDO, Base Lab, Chandigarh.*

1

Horticulture Importance and Scope

Pranava Pandey

German Prof. Peter Laurenberg conceived first the word horticulture. The term "Horticulture" is known in written form first in seventeenth century in a book entitled "New World of Words" by Phillips.

Etymologically the word **"Horticulture"** is derived from two Latin words viz. '**Hortus'** [garden or enclosure] and **'Cultura'** [cultivation]. So, literally horticulture means 'garden or enclosure culture' or 'cultivation of garden crops within protected enclosures'.

Garden: Garden is a broad term derived from the Latin word *Gyrdan* meaning "to enclose".

DIVISIONS AND DEFINITION OF HORTICULTURE

Pomology: (pomum = fruits, logy = science) It is a branch of horticulture which deals with study or cultivation of fruit crops starting from raising of saplings and intercultural to harvesting operations. The 'pomology' term is derived from latin word **pomum (**means fruits) and greek word **logy (**means science).

"**Poma**" in greek means fruits later subsequently transfer in to **Pome** in Latin word means fruits, logos-study.

Olericulture: Olericulture term is derived from two Latin words i.e. **Oleris** meaning Pot herb or ole (cabbage) and **Cultra** meaning cultivation. So, Olericulture literally means pot herb cultivation or the cultivation of vegetables.

Floriculture and Landscaping: It is a branch of Horticulture which deals with commercial cultivation, marketing and arranging the flowers and ornamental plants, which includes annuals, biennials and perennials i.e., trees, shrubs, climbers and herbaceous perennials for aesthetic purposes.

Landscaping: Landscaping means the design and alteration of a portion of land through the use of planting materials and land reconstruction.

Post Harvest Technology (PHT): PHT deals with post harvest handling, grading, packaging, storage, processing, preservation, value addition, marketing etc. of horticulture crops. In nut shell you can say it is the management of produce after harvest.

However, over the years the scope of the above field has been expanded to include other crops like plantation crops (Coffee, Tea, Rubber, Coconut, and Cocoa etc.), bamboo and mushroom. Bee keeping, one of the tools to improve the productivity of horticultural crops through enhanced pollination is also being taken care by the horticulture discipline. In view of the above developments Horticulture can now be redefined as the "Science of growing, improvement and management of fruits, vegetables, ornamentals, medicinal and aromatic crops, spices, plantation crops and their processing, value addition and marketing".

Horticulture as a science

Horticulture involves the application of fundamental sciences viz. physics, chemistry as well as plant sciences viz. biochemistry, plant physiology, botany, genetics and plant breeding etc.

Horticulture as an art

Landscape architecture, designing of different types of garden, edge, hedge and topiary with myriad species of plants, preparation of shrubbery and herbaceous borders by growing of different contrast colors of flowers as per need and climatic condition, propagating, spraying etc. required specialized techniques are proving the artistic scope of the horticulture. The ability attained through practice is skills develop into an art.

IMPORTANCE OF HORTICULTURE

India has witnessed increase in horticulture production over the last few years. Significant progress has been made in area expansion resulting in higher production. Over the last decade, the area under horticulture grew by 2.6% per annum and annual production increased by 4.8%. During 2017-18, the production of horticulture crops was 311.71 million tonnes (mt) from an area of 25.43 million hectares (mha). Fruit crops cover an area of 6.51 mha while vegetable crops occupy 10.26 mha with annual production of 97.36 mt and 184.40 mt, respectively. In our country per capita per day consumption of fruits is the lowest (120 g) in comparison to developed (200–350 g) and even the developing countries (100 –200 g). It is estimated that per capita fruits availability in our country is 141 g per day which is far below the recommended quantity of 230 g per capita per day. Whereas, per capita availability of vegetables is 272 g against recommended allowance of 300 g.

To meet out the demand of huge population of the country, horticultural production would be required in vast amount. Therefore, it is urgent need to increase the production and productivity of horticultural crops. The requirements of processing industry and foreign exchange further include the requirements of horticultural produce. In view of the above, there is lot of scope of increasing production and potentiality of horticulture crops.

Besides fruits and vegetables, floriculture industry in India comprising of florist trade, nursery plants, potted plants, seed and bulb products is being observed as sunrise industry. Area under loose flowers (marigold, jasmine, etc.) and cut flowers (rose, chrysanthemum, gladiolus, carnation, etc.) increasing day by day but loose flowers are grown on a commercial scale in the country.

Plantation crops, Spices, Medicinal and Aromatic plants are another potential sector with lot of opportunities for employment generation, foreign exchange earnings and overall supporting livelihood sustenance of mankind at large.

Horticulture has emerged as a vital part of agriculture, offering a wide range of choices to the farmers for crop diversification. It also provides ample opportunities for sustaining large number of agro-industries which generate substantial employment opportunities.

The parts like stem, leaf, flowers, roots and even the fruits of horticulture plants are used to make drugs, chemicals, insecticides, germicides etc. e.g. rose water is used to cure eyes ailments. Similarly saffron is important ingredient of many medicines.

As compared to field crops Horticultural crops give more returns per unit area (More yield in terms of weight and money).

Horticultural plants improve environment by reducing pollution, conserves soil and water and improve socio-economic status of the farmers.

In short and sweet horticulture supplies quality protective food for health and mind, more calories per unit area, develops better resources and higher returns per unit area. It also enhances land value and creates better purchasing power for those who are engaged in this industry. Therefore, you can say that **"horticulture is important for health, wealth, hygiene and happiness".**

The horticulture sector contributes around 31 % of the GDP from about 14 % of the area and 38 % of the total exports of agricultural commodities.

Nutritional importance of horticultural crops

Our body requires proteins, fats, carbohydrates, vitamins and minerals for its health. Fruits and vegetables play an important role in balanced diet as these crops provide high amount of vitamins and minerals to all of us. Proteins, fats, carbohydrates, vitamins and minerals etc. are recognized as protective foods as they are necessary for the maintenance of human health.

Essential nutrients and minerals provided by different fruits and vegetables are:

Vitamins	Role and Deficiency problems	Sources
A	a) Plays an important role in bone growth, tooth development, reproduction, cell division, and gene expression. b) Helping the eyes adjust to light changes Deficiency- night blindness, hamper growth, rushes in skin.	Plants contain carotenoids, some of which are precursors for vitamin A (e.g., α-carotene and β-carotene). Yellow and orange vegetables and fruits contain significant quantities of carotenoids. Green vegetables also contain carotenoids. Eg. Mango, Papaya, Persimon, Dates, Jack fruit, Oranges, Passion fruit, Coriander leaves, Drumstic leaves, Fenugreek leaves etc.
B_1	a) For normal functioning of digestive system and nervous system. Deficiency-Beri-Beri, loss of appetite and paralysis.	Legumes (e.g., beans) Walnut, Apricot, Apple, Banana, Grapefruit, Plum and Almond etc. Chillies, Colocasia leaves, Tomato, etc. are rich sources of thiamin.
B_2	a) For mitochondrial electron transport in body. b) Important for respiration and skin glow. Deficiency- sore throat, cataract.	Bael, Papaya, Litchi, leafy vegetables etc.
B_3 (Niacin)	Essential for energy production in all cells & DNA repair. Deficiency- Pellagra	Green leafy vegetables, coffee, tea etc.
B_5	A vital coenzyme in numerous chemical reactions. Play role in energy metabolism. Deficiency-Depression	avocados, cashew nuts, broccoli and peanuts
B_6 (Pyridoxin)	Has the most importance in human metabolism. Deficiency- skin disease	Banana, walnut, almond, apricot, apple, plum etc.
C	A highly effective antioxidant, acting to lessen oxidative stress. Vitamin C is sensitive to air, heat, and water, so it can easily be destroyed by prolonged storage, overcooking, and processing of foods. Deficiency- Scurvy, gum and tooth decay, delay in wound healing and rheumatism.	Barbados cherry, Aonla, Guava, Lime, Lemon, Sweet oranges, Chillies, Tomato, Coriander leaves, Drumstick leaves etc.
D	Plays a critical role in the body's use of calcium & phosphorous. It increases the amount of calcium absorbed from the small intestine and helps form and maintain bones. Defiency- rickets, tooth decay	body obtains vitamin D through the skin, which makes vitamin D in response to sunlight.
E	To maintain the integrity of the body's intracellular membrane by protecting	leafy vegetables

	its physical stability and providing a defense line against tissue damage caused by oxidation. Defiency-Fertility disorder	
K	Essential for the functioning of several proteins involved in blood clotting. Deficiency-Problem in blood clotting	leafy vegetables
Fibres	Increase water intake and digestibility	Banana, Guava, Pomegranate, Amaranth, Beet leaf, Spinach etc.
Proteins	For repair and maintenance of body tissues	Fruits and vegetables are low in proteins except Cashew nut Walnut, Almond, Avocado, Peas, cowpea, Bean etc.
Fats & Carbohydrates	Main source of Energy	Walnut, Almond, Avocado Banana, Dates, sweet potato, potato etc.

Minerals and their role:

Minerals	Role and Deficiency problems	Sources
Calcium	About 99% of the calcium in the body is found in bones and teeth, while the other 1% is found in the blood and soft tissue. The physiological functions of calcium are so vital to survival that the body will demineralize bone to maintain normal blood calcium levels when calcium intake is inadequate. Thus, adequate dietary calcium is a critical factor in maintaining a healthy skeleton.	Litchi, Fig, Phalsa, Citrus, Sapota, leaves of Drumstick , Curry leaves, Amaranthus, Radish leaves, Fenugreek etc.
Phosphorous	Essential for normal functioning of body.	Walnut, Avocado, Dates, Pomegranate and Grape raisins, Carrot, Drumstick leaves, Beans, etc.
Iron	Required for production of hemoglobin (oxygen carrier).Deficiency-anemia,	Karonda, Date palm, Grape raisins, West Indian Cherry, Guava, Sitaphal, Avocado, Sapota, plum etc. Amarantus tender, Coriander leaves etc.

SCOPE OF HORTICULTURE

India is bestowed with diverse climatic and edaphic conditions provides sufficient opportunity to grow a variety of horticulture crops. After having achieved the self sufficiency in food security, nutritional security for the people of the country has become a challenging task for horticulturists of the hour. To meet the nutritional requirement in terms of vitamins and minerals, horticulture crops are to be grown in sufficient quantities.

Area of arable land and holdings diminishing day by day by field crops, industry, housing, roads and infrastructure due to population explosion. Thus, available wasteland can be best utilized to cultivate hardy horticultural crops like fruits, plantation crops and medicinal and aromatic plants.

At present our share in international trade of horticultural commodities is less than one per cent of total trade. Moreover, these commodities (spices, coffee, tea etc.,) fetch 10-20 times more foreign exchange per unit weight than cereals and therefore, taking advantage of globalization of trade, nearness of big market and the size of production, our country should greatly involve in international trade which would provide scope for growth. There is enormous scope for export of fresh and processed products.

Development of financial institutions, co-operatives in rural areas helps farmer to replace their traditional crops with Horticultural crops that add more returns from these crops.

Aesthetic value and religious importance is unique feature distinguishing it from agricultural activities. Mango leaves, wood, banana leaves etc. are used for religious functions. Horticultural trees also keep up the ecosphere and protect the environment.

References

Anonymous, 2018. Horticultural Statistics at a Glance 2018. Horticulture Statistics Division Department of Agriculture, Cooperation & Farmers' Welfare Ministry of Agriculture & Farmers' Welfare, Government of India.

Chadha, K.L. 2001. Text book of Horticulture, ICAR, New Delhi.

Edmond, J.B., Sen., T.L., Andrews, F.S. and Halfacre, R.G. 1963. Fundamentals of Horticulture, Tata McGraw Hill Publishing Co., New Delhi.

Garner, V.R., Bradford, F.C. and Hooker, Jr. H.D. 1957. Fundamentals of Fruit Production, McGraw Hill Book Co., New York.

https://ecourses.icar.gov.in

https://www.agriglance.com/2016/10/31/importance-and-scope-of-horticulture

Kumar, N. 1990. Introduction to Horticulture, Rajyalakshmi Publications, Nagercoil, Tamil Nadu.

Naik, B.H. and Thippesh, D. 2014. Fundamentals of Horticulture and Production Technology of Fruit Crops, UAHS, Shimoga, India

Srivastava, R.P. and Kumar, S. 2001. Fruit and Vegetable Preservation: Principles and Practices. International Book Distributing Co., Lucknow, India.

Yildiz, F. 2010. Advances in Food Biochemistry, CRC Press, Taylor & Francis Group, US.

2

Novel Approaches for Sustainable Development of Horticulture

Sanjeev Kumar, Alka Singh, T.R. Ahlawat, Shailendra K. Dwivedi and Dushyant D. Champaneri

Horticulture has emerged as one of the potential agricultural enterprises in accelerating the growth of economy in India. It offers not only a wide range of options to the farmers for crop diversification, but also provides ample scope for sustaining large number of agro-industries which generate huge employment opportunities. Horticulture has been evolving in form of horticulture industry in our country owing to its significant role in nutritional security, poverty alleviation and employment generation. The production of fruits, vegetables and flowers has acquired much importance in recent times due to their increasing demand. India has a wide variety of climate and soil on which different horticultural crops such as fruits, vegetables, ornamentals, medicinal and aromatic plants, plantation crops and spices are cultivated successfully.

Horticultural crops could serve an ideal way of achieving sustainability in small holdings, increasing employment, improving environment, providing an enormous export potential and above all achieving nutritional security. Despite these advantages, India's share in the global market is insignificant accounting for only 1.7% of the global trade in vegetables and 0.5% in fruits and less than 0.5% in flowers. Hence, continuous efforts are required for strengthening the horticulture sector through novel sustainable updated and applied technologies. The major arenas as new dimensions to augment the horticulture sector are given below:

1. Quality planting material
2. Enhanced Nutrient Use Efficiency
3. Plasticulture and its alternatives

4. Technology Driven Farming and Farm Mechanization
5. Irrigation Systems
6. Interventions in Greenhouse Technology
7. Rejuvenation of Old Orchards
8. Precision Farming
9. Biotechnological Interventions in Horticulture
10. Hi-tech Interventions in Seed Spices
11. Pesticides Residue Management Strategies
12. Organic Horticulture
13. Horti-Tourism: Employment Opportunities for Youths
14. Post-Harvest Loss Management Strategies

1. Quality planting material

The planting material forms the base of an orchard, farm or greenhouse production system and is the key to deliver high quality and quantity in produce. Hence, the genuine quality in terms of genetic purity, healthy and disease and pest free plant material, be it seed or vegetative propagules needs to be assured. Source of planting material *i.e.* mother plant as well as rooting/ growing media and techniques needs to be standardized. High priority should be laid towards development of sound scientific nursery network in the country.

Some of the problems encountered during production of quality planting materials are:

i) Absence of genetically uniform rootstocks
ii) Lack of variability in rootstocks
iii) Less availability of information on rootstock-scion interactions
iv) Lack of facilities for round the year production through special structures
v) Lack of tissue culture protocols for many crops
vi) Lack of vegetative propagation techniques for seed propagated crops

There is also a need to start rootstock-breeding programme. Production of healthy and genuine planting material should be intensified in major commercial crops for the establishment of a sound horticulture industry. Accreditation of nurseries and tissue culture laboratories should be carried out in a systematic manner to ensure production and sale of quality planting material. In this context, National Certification System for Tissue Culture Raised Plants (NCS-TCP) was established in 2006, authorizing the Department of Biotechnology, Ministry of

Science & Technology as the Certification Agency under the "Seeds Act, 1966" to fulfill the very objective of production and distribution of quality planting material.

Emerging trends for quality planting material production

1. Soilless rooting/growing Media
2. Plug trays
3. Modification in grafting techniques for fruits, vegetables and flowers
4. Dwarfing rootstocks (HDP)
5. Mechanical substrate mixing, media filling and seed sowing

Important considerations for development of quality planting material

- Grafting in vegetables should be followed to mitigate biotic and abiotic stresses under open as well as protected conditions and thus minimizing use of plant protection chemicals.
- In citrus, micro-grafted/ micro-budded indexed plants should be made available on the pattern of Nagpur mandarin being produced from ICAR-CCRI (Central Citrus Research Institute), Nagpur. Similarly, micro-propagules should also be made available to growers in crops like banana, pomegranate, strawberry and potato. Public Private Sector (PPS) interventions are required in curtailing the cost of tissue culture (TC) plants in date palm and oil palm for the benefit of the growers.
- Indigenously developed diagnostic kits should be used for the management of viral diseases.
- Plug tray nursery production should be promoted in vegetable and flower crops.
- Enhancing the capacity of existing micro-propagation units in *Public-Private Partnerships* (PPP) mode.
- Establishing primary and secondary hardening facilities at zonal level to minimize the cost of TC plantlets.
- Strengthening of rootstock breeding in fruit and vegetable crops.
- High genetic purity of breeder, foundation and certified seeds of annual horticultural crops must be ensured.
- Production of F_1 hybrid seed of high-valued crops under protected structures.
- Production of potato minitubers in soilless culture (Aeroponics). The technique has already been standardized by CPRI (Central Potato Research

Institute), Shimla that provides multiplication rate of 5-8 times higher in terms of number of tubers produced as compared to traditional method.

2. Enhanced Nutrient Use Efficiency

Nutrient use efficiency (NUE) is a measure of how well plants use the available mineral nutrients. It can be defined as yield (biomass) per unit input (fertilizer, nutrient content). NUE is a complex trait that not only depends on the ability of a plant to take up the nutrients from the soil, but also on transport, storage, mobilization, usage within the plant and even on the environment. NUE is an important contributor to growth control and yield, the same levels of nutrients may cause growth and yield penalty in one species or variety but not in another one.

Improving nutrient use efficiency (NUE) is a fundamental challenge faced by the fertilizer industry in general and horticulture in particular. There is a need of integrated, balanced and effective nutrient management for the growth of horticultural sector. Nutrient Stewardship approach provides a framework for using the right nutrient source, applied at the right rate, at the right time, in the right place, to achieve improved sustainability in horticulture production. Different available technologies can be used to improve NUE as depicted in the figure below:

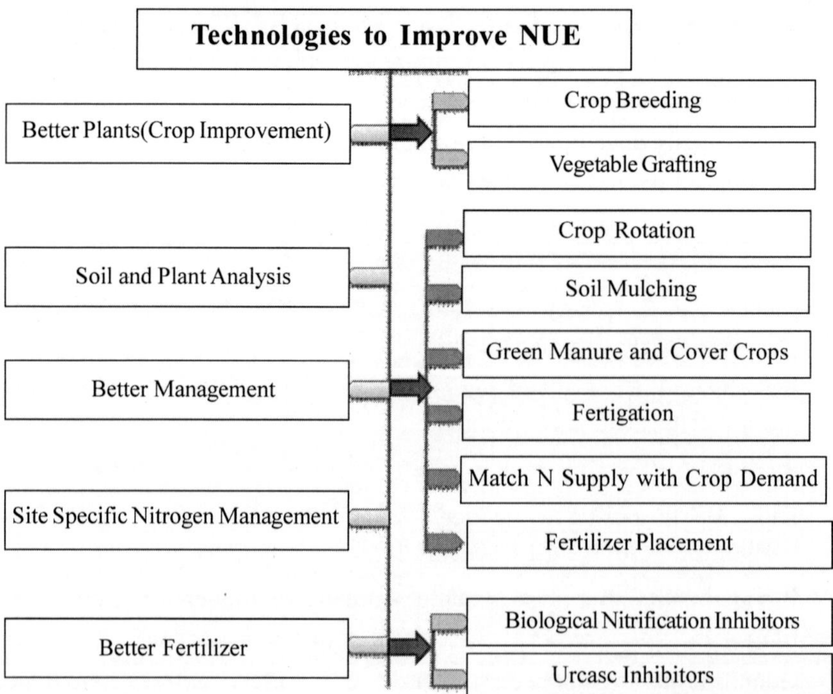

3. Plasticulture and its alternative

Plasticulture refers to the practice of using different plastic materials in agricultural operations/ applications like mulching, drip irrigation, row covers, low/high grow-tunnels, protected structure (low-medium-high tech greenhouses/shade net houses), silage bags, plug trays and pots. Use of plastics in horticultural crop production has increased dramatically in the last two decade in the following manners:

1. Plastic reservoirs and irrigation systems (water harvesting ponds, drip lines, sprinklers, *etc.*)
2. Mulching
3. Plastic for raising planting material, floating row cover, multi-celled plastic plug trays
4. Protected cultivation (Polyhouse and Low Tunnel polyhouse, shade net house *etc.*)
6. Post-harvest handling (Boxes, crates for crop handling, packaging and transportation)

Application of plasticulture is undoubtedly highly beneficial in horticulture however non-degradable nature of plastic has some ecological concerns. So, availability of environment friendly material for polyhouse cladding as well as reuse and recycling of used horticultural plastic waste and design of equipment and system for precise application of water, fertilizers, agro-chemicals to meet crop requirement is urgently needed as promotional policies as well as awareness programmes. The biodegradable plastic is made up of natural starch and vegetable oil derivatives especially of potato and tapioca starch which has been setting a new trend in general and new mode of value addition in horticulture produce.

4. Technology Driven Farming and Farm Mechanization

There is a need of technology driven development for small and marginal farmers with successful entrepreneurial models. Farm mechanization facilitates are very important for timely, precise and scientific farm operations for increasing farm input and labour use efficiency.

Strengthening of horticulture through technology driven farming

- Adoption of crop regulation strategies using growth regulators, pruning and maintenance of ideal canopy architecture *etc.*
- Promotion of high-tech horticulture practices for efficient production, better input use efficiency, and enhancement in quality, better packaging and efficient retail chain management for better returns.

- Rejuvenation of old orchards of fruits crops like mango, cashew, sapota *etc.*
- Adoption of High Density Planting (HDP) in mango, litchi, sweet orange, guava etc. for profitable cropping, high regular yields with higher productivity.
- Pollination management in horticultural crops, potential of which can be visualized from the following table.

S.No.	Crop	Yield gain (%) via pollinators mediation
1.	Apple	19.01
2.	Mango	3400
3.	Guava	21.25-108.30
4.	Citrus	11.68
5.	Lemon	57.10
6.	Litchi	28.55-123.77
7.	Pear	540.74
8.	Peach	116
9.	Cherry	283
10.	Cucumber	23-56
11.	Bitter gourd	24.10
12.	Ridge gourd	249.10
13.	Onion	470-1285
14.	Coriander	68-78

- Promotion of spice crops based framing systems to increase production and productivity.
- Promotion and adoption of GAP strategies in horticultural crops.
- Adoption of organized strategy for onion and potato production and sufficient storage for better distribution and price stabilization.
- Exploring the potential of colour pigments lycopene, â-carotene, anthocyanin, lutein *etc.* of fruits and vegetables for nutraceutical applications.
- Cultivation of microgreens: Microgreens are the new class of edibles, a very specific type which includes seedlings of edible vegetables, herbs or other plants. They are highly concentrated with various bio-active compounds like vitamins, minerals, antioxidants *etc.* compared to their mature counterparts, thus present a way forward to manage nutritional deficiencies being observed in one and other sections of population particularly in children of our country.

Key advantages of farm mechanization

- Increasing crop intensity and yield for better returns.
- Reducing the risk of weather and non-availability of labour.
- Improving working conditions and enhancing the safety of farmers.
- Converting uncultivable land for agricultural use through advanced tilling technologies.
- Increasing employment opportunities for skilled youth.

5. Irrigation Systems

Water is a vital component for crop management. The quality and amount of irrigation water is very critical and governed by crop type, soil type/media, weather conditions, environment control system *etc*. Drip and sub-irrigation systems are the most efficient means of irrigation over other systems of irrigation. Adoption of such systems not only results in enhanced water use efficiency but also minimizes the environmental pollution. Though, its potential benefits have well been established at research as well as in farmer fields and simultaneously, Government is also promoting it through subsidies, but its adoption rate is very low by the end users. Reduction in water consumption due to drip irrigation over the surface method of irrigation varies from 30 to 70% and productivity gain in the range of 20 to 90% for different vegetable crops. It also helps to reduce labour and electricity requirement, and improve fertilizers use efficiency over the conventional method of irrigation.

6. Interventions in Greenhouse Technology

Greenhouse technology modifies/or controls partially or fully micro-climate surrounding the plants as per the requirements with the added advantage of assured remuneration on account of quality produce and in some cases off-season/early production and prolonged production.

Several novelties and innovations under protected horticulture are discussed hereunder

i) *Ladakhi* polyhouse and trench cultivation for off-season vegetable production in cold deserts of Ladakh (J&K) and Lahaul-Spiti (HP)

ii) Rain shelters of different size and claddings in Kerala and North-East Regions for cultivation of vegetables like cauliflower, cabbage, broccoli and melons.

iii) Soil-less culture for the production of nursery plants, vegetables, flowers and certain fruits like strawberry

iv) Vertical farming of vegetables, fruit crops like strawberry and some of the annual flowers) hydroponics under natural or LED light

v) Edible seedlings/microgreens - new generation smart and immune booster edible greens

vi) Plasticulture applications as irrigation, fertigation, mulching, cladding materials, shade nets, containers, crop covers *etc.*

vii) Production of tissue culture plants in crops like banana, carnation, gerbera, orchids, chrysanthemum and other herbaceous crops.

viii) Vegetable grafts for developing biotic and abiotic stress resilient plant types in tomato, brinjal, cucurbits *etc.* as well as production of wonder plant like 'Pomato'.

ix) High quality seed production under protected structures in different vegetables through structural exclusion.

7. Rejuvenation of Old Orchards

Most of the orchards often encounter decline in productivity and poor fruit quality after certain years of plantation. Senile orchards with poor efficiency are now a common phenomenon in many fruit crops and have become uneconomic. They are characterized by inter-mingling and overcrowding of shoots with poor photosynthetic efficiency. The most of the problems are due to unsuitable site and climate, cultivation of intercrops, inadequate nutrition, improper planting, undesirable planting materials, incidence of insect-pests and diseases, and other biotic & abiotic problems.

As most of the fruit crops have long juvenile phase, instead of new plantations, it is advisable to rejuvenate these trees for better and quality production. Plant rejuvenation refers to the reversal of the adult phase in plants and the recovery of part or all of juvenile plant characteristics. The growth and reproductive vitality of plants can be increased after rejuvenation. Various steps involved in rejuvenation technology are dependent on orchard condition, age of the plant and canopy management. Rejuvenation is thus gaining popularity to make such orchards productive and has helped in restoring the productivity and profitability of several fruit crops like mango, guava, aonla, litchi, cashew in the shortest possible period besides reducing the tree height to a manageable limit.

8. Precision Farming

Precision Farming is a management strategy that employs detailed, site-specific information to manage production inputs precisely. It is basically, adding the right amount of input at the right time and the right location within a field. It's all about managing specific sites following the fundamentals of 3 Rs.

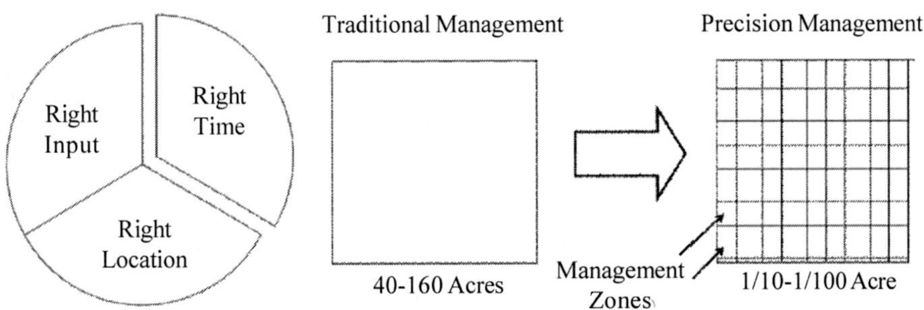

Right
Input
Right
Time
Right
Location
Traditional Management
40-160 Acres
Precision Management
Management
Zones
1/10-1/100 Acre

Precision Farming and its approach

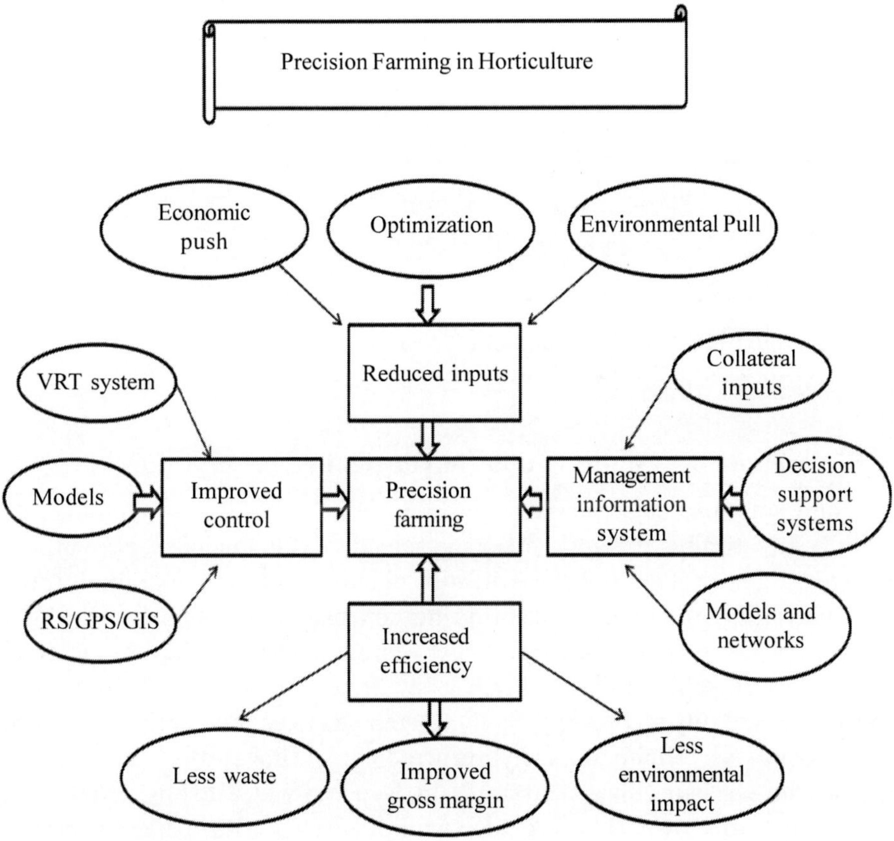

Precision Farming in Horticulture
Economic
push
Optimization
Environmental Pull
Reduced inputs
VRT system
Collateral
inputs
Models
Improved
control
Precision
farming
Management
information
system
Decision
support
systems
RS/GPS/GIS
Increased
efficiency
Models and
networks
Less waste
Improved
gross margin
Less
environmental
impact

Basic steps in precision farming

- Assessing variation (Information Generation)
- Managing variation (Technology Management)
- Evaluation

Components of precision farming (See chapter 15 for detailed description)

A. **Information or data base:** Soil, Crop, Climate

B. **Technology:** Computer System, Global Positioning System (GPS), Differential Global Positioning System (DGPS), Geographic Information System (GIS), Remote Sensing (RS), Sensors, Variable Rate Applicator, Yield Mapping Technology.

Important Applications of Precision Farming

1. Water Management (drip irrigation, sprinkler irrigation and fertigation)
2. Surface Covered Cultivation (mulching, soil solarization)
3. Controlled Environmental Condition (polyhouse, grow tunnels, shade net houses *etc*.)
4. Organic Farming
5. Precise Space Utilization (HDP, Meadow Orcharding)
6. Micro-Propagation (7) Integrated Pest Management (IPM)
8. Integrated Nutrient Management (INM)
9. Vegetable grafting

9. Biotechnological Interventions in Horticulture

The desirable molecular markers can be used in a breeding programme for identifying desirable parents, enhancing selection of elite alleles, pyramiding favourable alleles at multiple loci affecting either a single trait or multiple traits and fingerprinting cultivars for identification and plant variety legislation (PVP). Molecular markers linked with reproductive traits such as sex forms (cucurbits and papaya), cytoplasmic male sterility (sweet pepper, cole crops and carrot), self-incompatibility (cole crops) facilitates rapid conversion of promising lines into desirable sex form for use in hybrid breeding. The linked molecular markers to genes for resistance in tomato (leaf curl, bacterial wilt, late blight, root knot nematode), cauliflower (black rot, down mildew), cucumber (papaya ring spot virus, Fusarium wilt, downy mildew), banana (Fusarium wilt, browning nematode), potato (cyst & golden nematodes), grape (powdery mildew), apple (powdery mildew, scab), rose (black spot) and coconut (mites and lethal yellowing), and quality traits such as carotenoids and anthocyanins (carrot and cauliflower); lycopene (papaya and watermelon); pungency & sulphur assimilation (onion), glucosinolates (cole crops) and reduction of antinuritional factors such as tannins (broad bean); fruit acidity (citrus); and seedlessness (grape and cucumber) will have great promise in marker-assisted selection.

Further, genetic engineering has proved to be a reliable tool for manipulating tree architecture by altering bio-chemical pathways in fruit trees. Genetic modification of apple rootstock M26 using rolA gene has made it possible to reduce the shoot growth without changing rooting performance.

Biotechnological tools are also being explored for development nutraceutical varieties in fruits and vegetables. The use of molecular approach in recent time has made it possible to introgress desirable gene from plant sources in lieu of public disinclination about the use of GE food and next generation protein rich potato using AmA1 has been developed at CPRI, Shimla. Similarly, RNAi technology has established its importance for blocking and regulating bio-synthetic pathway of anti-nutritional factor in cassava. Metabolomics could be the futuristic approach for targeted metabolic profiling to identify, enrich and regulate biological compounds in most inclusive way.

10. Hi-tech Interventions in Seed Spices

Seed spices have low nutrient and water requirements and provide higher remuneration than cereals, pulses and oilseeds, thus playing a very important role in Indian economy. The average productivity of major seed spices (cumin, coriander, fennel and fenugreek) is only 9.8 q/ha, which needs to be enhanced to the level of 15.20 q/ha by 2050 to meet domestic as well as export demand. Looking to the vulnerability of these crops insect-pests and diseases, it seems to be a great challenge to ridge the wide gap between present and targeted productivity. Protected cultivation technologies, automatic fertigation, raised-bed planting techniques, plug-tray seedling production and transplanting technologies, insect-proof net covered walk in tunnel, plastic mulching and shade net could serve as hi-tech interventions for raising productivity of seed spices. Application of water and nutrient through fertigation and sowing of crops with raised-bed technique results in enhancing 30- 35% yield, with saving of 25-30% nutrient and 45-50% water, which ultimately enhance profitability in seed spice cultivation.

11. Pesticides Residue Management Strategies

In recent years, occurrence of pesticide residues in horticultural commodities has caused serious rejection of our consignments at international trade. Lack of Good Agricultural Practices (GAP) recommendations and unregulated applications of pesticides could be the probable reason for such food safety related non-compliances. In most of the importing countries, the maximum residue limit (MRLs) has been set as 0.01 mg kg^{-1} by default. In the absence of

GAP recommendations, the pre-harvest interval (PHI) or the safe waiting period are unavailable for many of the pesticidal combinations. Thus, crops at harvest usually contain residues above the MRLs, which make them unfit for export. The Food Safety Standards Authority of India (FSSAI) has also started the implementation of the new MRLs (0.01 mg kg^{-1}) to check the sale of such non-compliant commodities within India.

Effective residue monitoring programmes should include the following points:

1. Product traceability from farm to consumer level
2. Implementation of GAP
3. Adherence to the PHI (Pre-harvest Interval) recommendations
4. Label claim and the use of residue-free agro-inputs from reliable sources.

An effective residue monitoring program requires the support of scientifically established methods for the sampling of produce and analysis of contaminants. The food safety traceability program implemented in case of table grapes is an excellent example, which is being replicated to resolve the issues of contaminants in other horticultural products. A comprehensive residue monitoring of pesticides, therefore promotes acceptance among consumers within and across the countries.

12. Organic Horticulture

Organic farming is a production system which avoids or largely excludes the use of synthetically compounded fertilizers, growth regulators and live stock feed additives. To the maximum feasible extent, organic farming systems rely on crop rotations, crop residues, animal manures, legumes, green manures, off farm organic wastes and aspects of biological pest control to maintain soil productivity and tilth to supply plant nutrients and to control insect-pests, diseases and weeds. Organic horticulture is the science and art of growing fruits, vegetables, flowers, or ornamental plants by following the essential principles of organic agriculture in soil building and conservation, pest management and varietal preservation. The organic production of horticultural crops has experienced some of the greatest growth in recent years. This growth is driven by both growers' desires for more sustainable systems and consumers' concerns for their and their family's health as well as for the ecological balance.

Organic farming has the potential to provide benefits in terms of environmental protection, conservation of non-renewable resources and improved food quality. Indigenous Technical Knowledge (ITK) in several parts of the country could be explored.

Sikkim has been officially declared as first organic state in India owing to 75000 ha of land converted into certified organic farms by following the guidelines prescribed by National Programme for Organic Production (NPOP). National Organic Farming Research Institute (NOFRI) has been established at Gangtok, Sikkim during 87th Annual General Meeting of Indian Council of Agricultural Research.

13. Horti-Tourism: Employment Opportunities for Youths

Rainbow revolution is an integral development programme of agriculture, horticulture, forestry, sugarcane, fishery, poultry and animal husbandry. It includes the amalgamation of all the other agricultural revolutions like Green Revolution (Food grains), White Revolution (Milk), Yellow Revolution (Oilseeds), Blue Revolution (Fisheries), Golden Revolution (Fruits), Silver Revolution (Eggs), Round Revolution (Potato), Pink Revolution (Meat), Grey Revolution (Fertilizers) *etc*. The agricultural policy of 2000 envisaged holistic development of Indian agriculture and aimed to achieve through Rainbow revolution. Indian agriculture is in a way, a victim of its own past success, especially the Green revolution as per the Economic survey of 2015-16. It suggested an Integral development programme to make the agricultural sustainability and Rainbow revolution as a concept was developed eventually.

Horti-tourism, indeed, adds a new dimension to galloping horticulture industry especially, to the countryside and hinterland horticulture in the country. On the other hand, growing tendency of people to take a break from the urban fatigue and visit the serene country sides for enjoying natural beauty and aesthetic pleasure has led to growth of Horti-tourism. Vast tourist destinations extend great opportunities for Horti-tourism endeavour, which in turn absorb energetic youth for the task. "Horti-Eco Tourism" is the symbiotic association of farming sector, tourism industry and farm business and the horticulture components play a vital role to secure a multifunctional, sustainable and competitive enterprise, maintain the landscape around the countryside, contribute to the vitality of rural communities, environmental protection, aesthetic pleasures *etc*. Eco-friendly plantations of horticultural crops like fruits and flowers of regional specialty.

From historical times India is regarded as the 'House of spices'. Coffee-based cropping systems comprising of spices like cardamom and pepper in Kodagu district of Karnataka, coconut and areca nut based cropping systems of Goa, Karnataka and Maharashtra, and tea based cropping systems (tea + pepper) are envisaged as potential Horti-eco-tourism destinations. Potential of Horti-tourism venture lies in its strength for multiple employment opportunities to the youth. However, comprehensive planning by the stakeholders for this venture opens up vast and multiple linkages within the niche domain to make this a reality through the empowerment of villages and the villagers. Horti-tourism centres can be the effective learning centres for school children where, children can see crops practically, understand their role in human health, and enjoy the local skills and products.

14. Management of Post-Harvest Losses

Post-harvest losses in different horticultural crops may occur from 6.70% (papaya) to 15.88% (guava), from 4.58% (tapioca) to 12.44% (tomato) in vegetable crops and 1.18% (black pepper) to 7.89% (sugarcane) in plantation crops. Harvesting, sorting/ grading, transportation, storage at wholesaler and retailer levels are the main operations and channels where losses are found to be high. It is very important to train farmers on Good Harvesting Practices (GHP) in order to minimize harvest and post-harvest losses. The following recommendations have emerged out to achieve the targeted goals:

- Skill development in all stakeholders for harvesting and handling.
- Infrastructures development such as construction of multi-commodity cold storage in rural areas and *mandies*.
- Setting up evaporative cooling based storage structures at farm and retail levels.
- Development of service sector like introduction of long distance transport using horti-train
- Establishment of horticulture crop processing training-cum-incubation centres, Horti- Processing Centres (HPC), Horti-parlour as Start-up Company.
- Incorporation of 'nano-emulsion', 'nano matrix' and 'nano-film' with smart packaging of horticultural commodities

References

Altman, A. 1999. Plant biotechnology in the 21st century: the challenges ahead. Electronic Journal of Biotechnology. 2 (2). http://www.ejb.org/content/vol2/issue2/full/1/.

Anonymous, 2007. Report of the Working Group on Horticulture, Plantation Crops and Organic Farming for the XI Five Year Plan.

Anonymous, 2011. Report of Planning Commission Working Group on Horticulture and Plantation Crops for XII Five Year Plan .

Anonymous, 2015. Raising Agricultural Productivity and Making Remunerative for Farmers. NITI Aayog, Government of India.

Barbuddhe, S.B. and Singh, N.P. 2014. Agro-Eco Tourism: A new dimension to agriculture. Technical Bulletin No. 46, ICAR Research Complex for Goa, Old Goa.

Bliss, F.A. 1999. Nutritional improvement of horticultural crops through plant breeding. HortScience. 34(7): 1163-1167.

Chan, K. 2016. Manual on Good Agricultural Practices (GAP). Asian Productivity Organization, Tokyo, Japan.

Das, A., Sau, S., Pandit, M.K. and Saha, K. 2018. A review on importance of pollinators in fruit and vegetable production and their collateral jeopardy from agro-chemicals. Journal of Entomology and Zoology Studies. 6(4): 1586-1591.

Evans, R.G. and Sadler, E.J. 2008. Methods and technologies to improve efficiency of water use. Water Resources Research. 44 (W00E04). doi:10.1029/2007WR006200.1.

Fageria, N.K. and Baligar, V.C. 2005. Enhancing nitrogen use efficiency in crop plants. *Advances in Agronomy.* 88: 97-185.

FAO, 2013. Good Agricultural Practices for greenhouse vegetable crops. Food and Agriculture Organization of the United Nations, Rome.

FAO, 2016. A scheme and training manual on Good Agricultural Practices (GAP) for fruits and vegetables (Vol. 1). The Scheme-standard and implementation infrastructure. Food and Agriculture Organization of the United Nations, Regional Office for Asia and the Pacific, Bankok.

Gopal Lal, 2018. Scenario, Importance and Prospects of Seed Spices: A Review. Current Investigations in Agriculture and Current Research. 4 (2): 491-498.

Gupta, H. 2013. Organic Farming and Horticulture: New Dimensions of Agriculture Development in MP, India. International Research Journal of Social Sciences. 2 (7): 14-18.

Haldhar, S.M. Kumar, R. Samadia, D.K., Singh, B. and Singh, H. 2018. Role of insect pollinators and pollinizers in arid and semi-arid horticultural crops. Journal of Agriculture and Ecology. 5: 1-25

https://agritech.tnau.ac.in/agriculture/agri_irrigationmgt_microirrigation.html

Jahangeer, A., Baba, P., Ishfaq Akbar and Vijai Kumar, 2011. Rejuvenation of old and senile orchards: A review. Annals of Horticulture. 4 (1): 37-44.

Jha, R.K. 2014. Kurukshetra- A Journal on Rural Development. 62 (8): 1-48.

Kuchi, V.S., Zehra Salma and Jahangir, K. 2017. Horti-Tourism: A value added approach for economically strengthening farmers. International Journal of Pure and Applied Bioscience. **5** (3): 793-795.

Malhotra, S.K. 2016. Research and development of seed spices for enhancement of productivity and profitability in the era of climate change. Indian Journal of Arecanut, Spices and Medicinal Plants. 18 (1): 3-18.

Mitra, S. and Hidangmayum Devi, 2016. Organic horticulture in India. Horticulturae. 2 (17). doi:10.3390/horticulturae2040017.

Nieves-Cordones, M, Rubio, F. and Santa-Maria, G.E. 2020. Nutrient use-efficiency in plants: An integrative approach. Frontiers of Plant Sciences. doi: 10.3389/fpls.2020.623976.

Pardeep Kumar, Parveen Sharma, Binny Vats and Vibhuti Sharma, 2018. Advances in vegetable grafting for mitigation of biotic and abiotic stresses in greenhouse vegetable production system. In: Technologies and Sustainability of Protected

Cultivation of Hi-Valued Vegetable Crops (Sanjeev Kumar, N.B. Patel, S.N. Saravaiya, and Patel, N.B., B.N. Patel, Eds.). Navsari Agricultural University, Navsari.

Patel, B.N., Alka Singh and Sanjeev Kumar, 2017. New Dimensions in Hi-Tech Horticulture. *In*: Approaches for Doubling Farmers' Income. (Patel, K.G., Virendra Singh, Shrivastava, A., Sangani, S.L., Dhandhukia, R.D., Deepa Hiremath and Patel, J.J., Eds.). pp. 403-417.

Rai, S.K., Arora, N., Pandey, N., Meena, R.P., Shah, K. and Rai, S.P. 2012. Nutraceutical enriched vegetables: Molecular approaches for crop improvement. International Journal of Pharma and Bio Sciences. 3(2). www.ijpbs.net.

Sanjeev Kumar, Lathiya Jasmin B. and Saravaiya, S.N. 2018. Microgreens: A new beginning towards nutrition and livelihood in urban-peri-urban and rural continuum. In: Technologies and Sustainability of Protected Cultivation of Hi-Valued Vegetable Crops (Sanjeev Kumar, N.B. Patel, S.N. Saravaiya, and Patel, N.B., B.N. Patel Eds.). Navsari Agricultural University, Navsari.

Sanjeev Kumar, Nikki Bharti and Saravaiya, S.N. 2018. Vegetable grafting: A surgical approach to combat biotic and abiotic stresses- A Review. Agricultural Reviews. 39(1): 1-11.

Sanjeev Kumar, Patel, B.N., Saravaiya, S.N. and Patel, N.B. (Eds.). 2018. Technologies and Sustainability of Protected Cultivation for Hi-Valued Vegetable Crops. Navsari Agricultural University, Navsari, Gujarat, India. 494p.

Sanjeev Kumar, Saravaiya, S.N. and Pandey, A.K. 2021. Precision Farming and Protected Cultivation: Concepts and Applications, Jaya Publishing House, Delhi.

Sarma, M.K., Baruah Sangeeta, Sharma, A.K. and Singh, B. 2017. Biotechnological interventions in Horticulture. Annals of Horticulture. 10 (2): 03-119.

Sarolia D.K., Samadia, D.K., Choudhary, B.R. and Singh, D. 2018. Production of quality seed and planting materials. CiAH/tech./Pub. No. 75. ICAR-Central institute for Arid Horticulture, Bikaner, Rajasthan.

Sharma, C., Dhiman, R., Rokana, N. and Panwar, H. Nanotechnology: An untapped resource for food packaging. Frontiers in Microbiology. doi: 10.3389/fmicb.2017.01735

Singh, H.P., Uma, S., Selvarajan, R. and Karihaloo, J.L. 2011. Micropropagation for production of quality banana planting material in Asia Pacific. Asia-Pacific Consortium on Agricultural Biotechnology (APCoAB), New Delhi, India.

Srivastava, A.K. and Malhotra, S.K. 2017. Nutrient use efficiency in perennial fruit crops: A Review. Journal of Plant Nutrition. Doi: 10.1080/01904167.2016.1249798.

Theofanis, A.G., Spyros, F., Katerina, A. 2011. Precision agriculture applications in horticultural crops in Greece and worldwide. In: Proceedings of the International Conference on Information and Communication Technologies for Sustainable Agri-production and Environment, Skiathos, September 8 (11): 2011.

Wittwer, S.H. 1993. World-wide use of plastics in horticultural production. HortTechnology. 3 (1): 1-14.

Wittwer, S.H. and Castilla, N. 1995. Protected cultivation of horticultural crops worldwide. HortTechnology. 5 (1): 1-17.

3

Horticultural and Botanical Classification of Plants

Pranava Pandey

HORTICULTURAL CLASSIFICATION OF PLANTS

The Horticultural crops are majorly classified on the basis of its consumption and uses as fruits, vegetables, flowers, spices, medicinal, aromatic and plantation crops. Moreover, these crops have been further classified into various groups on the basis of their habitat, growth pattern, edapho-climatic requirements and uses. All branches of horticulture classified specifically as hereunder:

A Classification of fruits

1. Based on Climatic adaptability

   a) **Tropical fruits:** These fruit crops are grown in tropical climatic regions which are geographically located between the parallels of 23° 27' South (Tropic of Capricorn) and 23° 27' North (Tropic of the Cancer) of the equator. They are generally evergreen and very sensitive to cold. Eg. Mango, Pineapple, Banana, Papaya, Citrus, Sapota, Avocado etc.

   b) **Sub-tropical fruits:** Fruit crops are grown in between temperate and tropical climatic regions. They may be either deciduous or evergreen nature. Eg. Mango, guava, citrus, durian, jackfruit, bread fruit etc.

   c) **Temperate fruits:** The temperate zones are mainly found between the tropics and the polar regions. Temperate fruit crops require distinctly cold and exposure of specific chilling temperature for certain period for flowering. They are generally deciduous in nature. Eg. apple, almond, peach, pear, plum, strawberry, apricot, persimmon, cherimoya, pecan nut, walnut, hassle nut, cherry, pistachios and kiwifruits etc.

2. **Based on rate of ethylene biosynthesis:** On the basis of amount of ethylene production fruits are divided in two categories as climacteric and

non-climacteric. Climacteric fruits produce much larger amount of ethylene than non climacteric fruits.

Climacteric fruits: Mango, Banana, Sapota, Guava, Papaya, Apple, Fig, Peach, Pear, Plum, Annona, Tomato etc.

Non-Climacteric fruits: Citrus, Grape, Pomegranate Pineapple Litchi, Ber, Jamun, Cashew, Cucumber, Cherry, Strawberry, etc.

3. Based on photoperiodic responses
   a) **Long day plants:** Passionfruit, Banana, Apple
   b) **Short day plants:** Strawberry, Pineapple, Coffee
   c) **Day neutral plants:** Papaya, Guava
4. Based on relative salt tolerance
   a) **High tolerant:** Datepalm, Ber, Amla, Guava, Coconut, Khirni
   b) **Medium tolerant:** Pomegranate, Cashew, Fig, Jamun, Phalsa
   c) **Sensitive:** Mango, Apple, Citrus, Pear, Strawberry
5. Based on relative acid tolerance
   a) **High tolerant:** Stawberry, Raspberry, Fig, Bael, Plum
   b) **Medium tolerant:** Pineapple, Avocado, Litchi
   c) **Sensitive:**
6. Based on morphology of fruits
   1. **Simple fruit – Berry**: Banana, Papaya, Grape, Sapota, Avocado etc.
   2. **Modified berry**

      Balusta - Pomegranate
      Amphisarca - Woodapple, Bael
      Pepo - Water melon
      Pome - Apple, Pear, Loquat
      Drupe - Mango, Pear, Plum
      Hesperidium - Citrus
      Nut - Cashew, Litchi, Walnut, Rambutan
      Capsule - Aonla, Carambola
   3. **Aggregate fruits**: Etario of berries-Custard apple, Raspberry etc.
   4. **Multiple fruit**: Syconus-Fig; Sorosis - Jackfruit, Pineapple, and Mulberry etc.

7. Based on longevity of fruits:
   a) **Very Long longevity (>100 yrs)**-Datepalm, Coconut, Arecanut
   b) **Long longevity (50-100 yrs)**-Mango, Tamarind
   c) **Medium longevity (10-50 yrs)**-Litchi, Guava, Pomegranate
   d) **Short longevity (<10 yrs)**-Pineapple, Banana

B. Classification of Vegetables

1. Based on the nature and morphology of plants
   a) Leafy, floral and succulents: Leafy vegetables- spinach, lettuce, celery, Cole crops, leeks, Green onion, asparagus, artichoke etc.
   b) Shrubs: Brinjal, chilli, tomato, etc.
   c) Trees: Drumstick, jackfruit, etc.
   d) Vines: Cucurbits, etc.
2. Based on the life span:
   a) Annuals: e.g. Tomato, brinjal, chilli, cabbage, cauliflower, cucurbits, leafy vegetables, etc.
   b) Biennials: Root and bulb crops e.g. Onion, radish, carrot, turnip etc.
   c) Perennials: e.g., Drumstick, Asparagus, Pointed gourd, etc.
3. Based on intercultural groups
   a) Perennial vegetables: Asparagus, Artichoke, Chow chow, Moringa, Ivy guard, Pointed gourd, Spine gourd
   b) Greens: Amaranthus, Spinach, Kale, Chard, Mustard, Collards
   c) Salad crops: Celery, Lettuce, Cress, Parsley
   d) Cole crops: Cabbage, Cauliflower, Broccoli, Brussel's sprout, Knol-khol
   e) Root crops: Beet root, Carrot, Parsnip, Turnip, Radish
   f) Bulb crops: Onion, Leek, Garlic, Shallot,Chive
   g) Tuber crops: Potato, Sweet potato, Cassava, Elephant foot yam
   h) Peas and beans: Pea, Bean, Broad bean, Lima bean, Winged bean, Cowpea
   i) Solanaceous crops: Tomato, Brinjal, Chilli, Pepper
   j) Cucurbits: Cucumber, Watermelon, Pumpkin, Gourds
   k) Fibre crop: Okra

l) Pot he**rbs green:** Spinach, Kale

m) **Other root crops:** Colocasia, Dioscorea, Arrow root

4. Based on climatic requirements

a) **Temperate vegetables:** Radish, potato, carrot, cole crops etc.

b) **Tropical and subtropical vegetables:** Cucurbits, tomato, brinjal, chilli, etc.

5. Based on growing season

a) ***Kharif*/Rainy season:** eg. Cucurbits, okra, beans etc.

b) ***Rabi*/Cool season:** eg. Cole crops, root crops, peas, leafy vegetables, etc.

c) **Summer/Warm season:** eg. Beans and Cucurbits

6. Based on plant part used as vegetable

a) **Stem:** Asparagus, Potato, Knol khol

b) **Leaves:** Spinach, Coriander, Amaranthus, Fenugreek

c) **Fruits:** Tomato, Brinjal, Okra, Peas, Beans, Watermelon, Pumpkin, Chillies

d) **Flower:** Cauliflower, Broccoli, Globe artichoke

7. Based on respiration rate of fruits and vegetables

Class	**Range at 5°C ($mg\ CO_2\ kg^{-1}\ hr^{-1}$)**	**Commodities**
Very low	< 5	Dates, Dried fruit and vegetables, Nuts, etc.
Low	5-10	Apple, Beet, Celery, Citrus Fruits, Garlic, Grapes, Kiwi Fruit, Onion, Papaya, Pineapple, Potato (Mature), Sweet Potato, Watermelon etc.
Moderate	10-20	Apricot, Banana, Cabbage, Carrot (Topped), Cherry, Fig, Lettuce (Head), Mango, Peach, Pear, Plum, Potato (Immature), Radish (Topped), Tomato, Summer squash
High	20-40	Avocado. Carrot (with tops), Cauliflower, Leeks, Lettuce (Leaf), Radish (with tops), Raspberry
Very High	40-60	Artichoke, Bean Sprouts, Broccoli, Brussels sprouts, Cut flowers, Green Onion, Okra
Extremely High	> 60	Asparagus, Mushroom, Parsley, Peas, Spinach, Sweet corn

C. Classification of Ornamental Plants

1. Based on growing season

i) Annuals

a) **Summer annuals:** Zinnia, Kochia, Portulaca, Tithonia, Gaillardia, Gomphrena, Sunflower, Daisy, etc.

b) **Rainy season annuals:** Balsam, cock's comb, Celosia, Gaillardia, etc.

c) **Winter annuals:** Antirrhinum, China aster, Cornflowers, Larkspur, Sweet Sultan, Phlox, Verbena, Candy tuft, Petunia, etc.

ii) **Perennials:** Rose, Jasmine, Crossandra, Orchids, Chrysanthemum, Berlaria, Hibiscus, Bulbous Crops, etc.

2. Based on colour of flowers

a) **White:** Antirrhinum. Alyssum, Dianthus, China aster, Zinnia, Chrysanthemum, Gladiolus, Gerbera, etc.

b) **Purple, Lavender or Blue:** Daisy, Rose, Carnation, Dahlia, Ageratum, China aster, Clitoria, Delphinium, Petunia, Viola, Verbena, Tithonia, Daisy, etc.

c) **Yellow and orange:** Antirrhinum, Marigold Calendula, Zinnia, Gaillardia, Rose, Gladiolus, Carnation, etc.

d) **Red pink:** Antirrhinum, Rose, Gladiolus, Carnation, Gerbera, Dahlia, etc.

3. Based on purpose of Growing

a) **Rockery:** Ageratum, Alyssum, Brachycome, Phlox, Portulaca, Linum, Nemesia, Saponaria, Godetia, Euphorbia, etc.

b) **Hanging basket:** Dwarf Ageratum, Petunia, Portulaca, Verbena, Torenia, Begonia, etc.

c) **Fragrant flowers:** Sweet Alyssum, Sweet Sultan, Sweet pea, Stock, Phlox, Carnation, Rose, Jasmine, Tuberose, etc.

d) **Bedding purpose:** Dahlia, Marigold, Phlox, Verbena, Carnation, Petunia, Ice Plant, Candy Tuft, Balsam, Portulaca, etc.

e) **For pots:** Carnation, Chrysanthemum, Dahlia, Rose, Antirrhinum, Petunia, Agloenema, Alocasia, Anthurium, Orchids, Aralia, Begonia, Chlorophytum, Dracena, etc.

f) **For Dry flowers:** Statice, Helichrysum, Acroclinum, Gomphrena, Limonium, Marigold, Rose, Lady's Lace, Nigella, etc.

g) **For Hedge purpose:** Lawsonia, Duranta, Tecoma, Bougainvillea, Thevetia, Hibiscus, Murraya, Dodonea, Acalypha, Aralia, Ipatorium, Chlorodendron spp, etc.

h) **For Edge purpose:** Dwarf Ageratum, Alyssum, Brachycome, Dianthus, Nigella, Portulaca, etc.

i) **For Cut flowers:** Rose, Chrysanthemum, Carnation, Gerbera, Anthurium, Orchids, Gladiolus, Tuberose, etc.

j) **For loose flowers:** Marigold, Jasmine, Crossandra, Berlaria, Chrysanthemums, China Asters, Sunflowers, Zinnia, Gaillardia, Rose, Dahlia, etc.

4. Based on purpose of use

a) Specimen purpose: Araucaria, Ficus spp. etc.

b) Flowering Trees: Gulmohar, Neelmobo, Cassua

c) Road Side Trees/ Avenue tree: Neem, Baniyan tree, Rain tree, Mahogany

d) Flowering Shrubs: Nerium, Hibiscus, Tagar

e) Foliage Shrubs: Thuja, Casurina

f) Climbers and Creepers: Petrea, Bignonia, Ipomea

g) Hedge and Edges: Duranta, Clearodendron, Ageratum

BOTANICAL CLASSIFICATION OF PLANTS

R.H. Whittaker (1969), a taxonomist was the first to propose the five kingdom taxonomic classification for living organisms on the basis of cell organization (prokaryotic or eukaryotic), body organization (unicellular or multicellular), mode of nutrition (holophytic, absorptive and ingestive), source of nutrition (autotrophic or heterotrophic), major ecological role (producer, consumer or decomposer) and phylogenetic relationship. These five kingdoms are Monera, Protista, Fungi, Plantae/Metaphyta (Plants), and Animalia/Metazoa (Animals).

The Kingdom Plantae includes all the plants. They are eukaryotic, multicellular and autotrophic organisms. Plant kingdom is classified as here under:

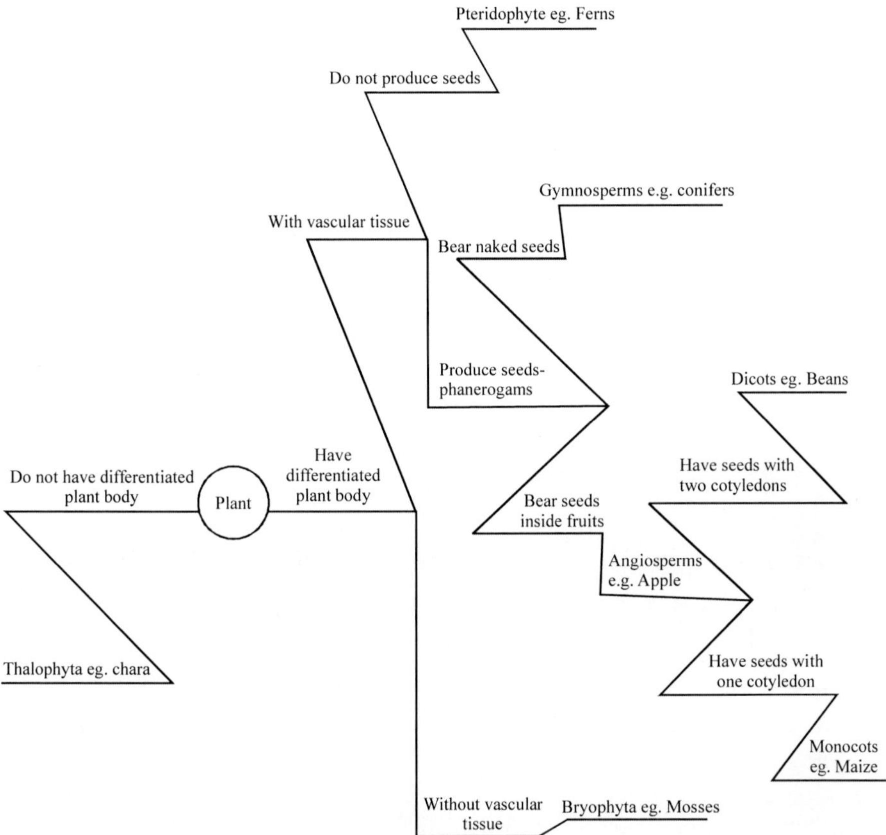

Plant Kingdom is divided into:

1. **Thallophyta:** Various types of microorganisms like algae, fungi and bacteria have been kept under it. Algae are Autotrophic in nature.
2. **Bryophytta:** Plants are found at land and water but are amphibians like Liver warts, Horn warts, Moss etc. These plants are also autotrophic as chloroplasts are present.
3. **Tracheophyta or vascular plants:** These plants have well developed vascular tissues and divided in to xylem and phloem. Further it is divided in to three subgroups: Pteridophyta, Gymnosperms and Angiosperm.
   a) **Pteridophyta:** In these plants there are lack of seeds and flowers. Examples: Club Mosses, horsetails, ferns etc.
   b) **Gymnosperm:** The plants whose seeds are completely uncoated and there is complete lack of ovary. Examples: Cycas, Pinus (Pines), Cedrus (Deodar) etc.

c) **Angiosperm:** This is the most- important subgroup of plants, whose seeds are coated and developed in an organ or ovary. Our major food, fibre, spice and beverage crops are flowering plants (angiosperms).

Further Angiosperm is classified into two categories

i) **Monocotyledonae (monocot):** Leaves of these plants are much longer rather than their width. Stems of monocot lack cambium and hence they increase little in girth except palm tree. Examples: Maize, wheat, rice, onion, sugarcane, barley, banana, coconut etc.

Characteristics

- Seed of these plants have one cotyledon.
- Their leaves have parallel venation.
- The roots of these plants are not developed.
- The flowers are trimerous i.e have three or multiple of three petals.
- In the vascular part, cambium doesn't exist.

ii) **Dicotyledonae (Dicot):** These plants have veins forming a network in their leaves. Almost have all the hardwood tree species, pulses, fruits, vegetables etc. Examples: Pea, potato, sunflower, rose, banyan, apple, neem etc.

Characteristics

- Seed of these plants have two cotyledons.
- In the vascular part cambium exists.
- The flower of the plant has multiples of four or five petals.
- These plants have secondary growth.

Botanical classification mainly dealt the identification, nomenclature and classification under taxonomy.

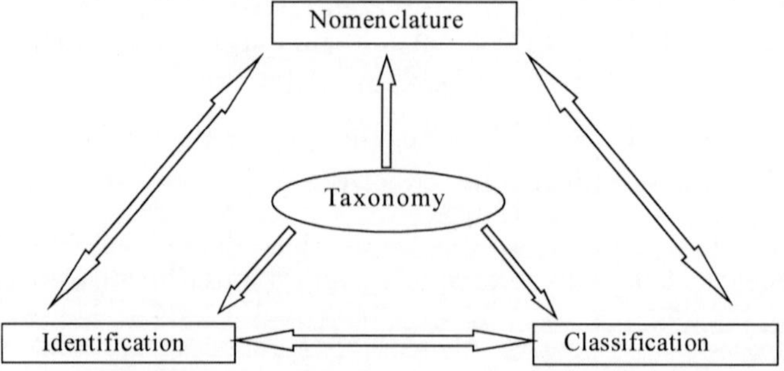

1. Natural populations: International Code of Botanical Nomenclature (ICBN)
2. Cultivated populations: International Code of Nomenclature for Cultivated Plants (ICNCP):

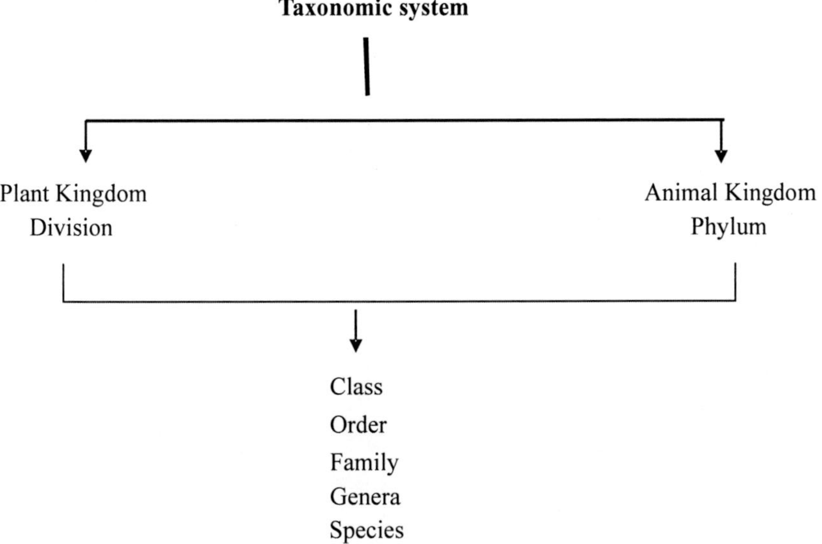

Class
Order
Family
Genera
Species

References

Arora, J.S. 2006. Introductory Ornamental Horticulture. Kalyani Publishers, Ludhiana.

Bose, T.K., Kabir, J., Maity, T.K., Parthasarathy, V.A. and Som, M.G. 2003. Vegetable Crops, Naya Udyog Publisherss, Kolkata.

Chadha, K.L. 2001. Text Book of Horticulture, ICAR, New Delhi.

Chattopadhyay, T.K. 1997. Text Book on Pomology (Fundamentals of fruit growing), Kalyani Publishers, Hyderabad.

Chundawat, B.S. 1990. Arid Fruit Culture, Oxford and IBH, New Delhi.

Edmond, J.B., Sen., T.L., Andrews, F.S. and Halfacre, R.G. 1963. Fundamentals of Horticulture, Tata McGraw Hill Publishing Co., New Delhi.

Garner, V.R., Bradford, F.C. and Hooker, Jr. H.D. 1957. Fundamentals of Fruit Production, McGraw Hill Book Co., New York.

https://ecourses.icar.gov.in

Kumar, N. 1990. Introduction to Horticulture, Rajyalakshmi Publications, Nagercoil, Tamil Nadu.

Randhawa, G.S. and Amitabha Mukhopdhyay, 2004. Floriculture in India, Randhawa, Allied Publishers Pvt. Ltd., New Delhi.

Simmonds, N.W. 1966. Banana, II Edition, Longman, London.

4

Seed Dormancy and Germination

Hetal Rathod, A.K. Pandey, Mital Dudhat and Madineni Tejaswini

A **seed** (in some plants, referred to as a **kernel**) is a small embryonic plant enclosed in a covering called the seed coat, usually with some stored food. It is the product of the ripened ovule of gymnosperm and angiosperm plants which occurs after fertilization and some growth within the mother plant. The formation of the seed completes the process of reproduction in seed plants (started with the development of flowers and pollination), with the embryo developed from the zygote and the seed coat from the integuments of the ovule. In broad sense, seed is a material which is used for planting or regeneration purpose. Seed technological point of view, seed may be sexually produced matured ovule consisting of an intact embryo, endosperm and or cotyledon with protective covering (seed coat). It also refers to propagating materials of healthy seedlings, tuber, bulbs, rhizome, roots, cuttings, sets, slips, all types of grafts and vegetatively propagating materials used for production purpose. Quality seed, therefore, can be defined as the seed which fulfills the minimum requirements of getentic and physical purity, seed germination and vigour freedom from weeds and other crop seeds and other relevant seed standards.

Types of seed

Orthodox: Orthodox seeds undergo a period of drying during maturation and shed at low water content. Such seeds can be dried down to a low moisture content of around 5% to 10% and successfully stored at low or sub freezing temperatures for long periods. These seeds acquire desiccation tolerance during development. The capability of seeds to withstand dehydration to is due to the presence of LEA (Late Embyogenesis Abundant) protein produced during the late phase of embryogenesis in presence of abscisic acid responsible for dessication tolerance.

Examples: Brinjal, chilli, tomato, okra, guava, sapota, banana, apple, cherry *etc.*

Recalcitrant: They do not go under drying at maturation and shed hydrated from the plant. Seeds which cannot survive drying below or relatively high moisture content (30-50%) hence cannot be stored for long periods. The sesiccation sensitivity of these seeds varies with the species. Water content in such seeds ranges from 30 to 80% on a wet mass basis.

Examples: Mango, litchi, avocado, citrus, jamun, jackfruit, mangosteen, durian, rambutan *etc.*

Dicotyledonous and Monocotyledouns seeds

On the basis of the number of cotyledons present in a seed, plants have been divided into two big classes: dicotyledons and monocotyledouns.

Monocotyledonous: Seed bears only one cotyledon in its embryo. The single cotyledon is terminal and the axis is lateral in position, e.g. Bamboo, banana, asparagus, ginger, tulip, lily, palms *etc.*

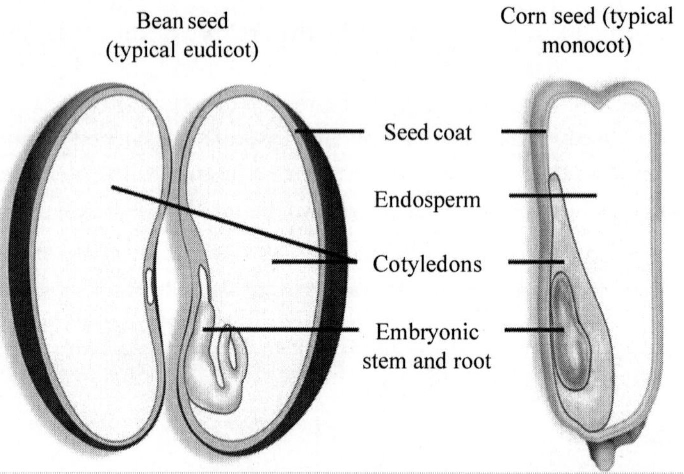

Dicotyledouns: Seed bears two cotyledons in its emryo. The axis is terminal and two cotyledons are lateral in position, e.g. gram, bean, pea, *etc.*

Albuminous and Ex-albuminous seeds: Seeds having a special food storage tissue, endosperm are known as albuminous or endospermic (*e.g.* Coconut) and the seeds that do not have such special tissue for food storage are known as ex-albuminous or non-endospermic. (*e.g.* bean, pea, walnut, squash, radish *etc.*)

Seed dormancy: Dormancy is a condition where seeds of several fruit and vegetable crops do not germinate even under favourable environmental conditions of water, temperature and air. Most of the fruit crop seeds require more time for germination and grow poorly. The dormancy in seed may be due to presence of hard seed coat, germination inhibitors, immature embryo, impermeable seed coat to water and oxygen, presence of high concentrate solutes. Such seeds should be exposed to special treatments like scarification, stratification, water soaking, growth regulator etc. to break the dormancy.

It has been observed that seeds of some fruit plants (mango, citrus) germinate immediately after extraction from the fruit under favourable conditions.

However, in others (apple, pear, cherry) germination does not take place even under favourable conditions. This phenomenon is called as dormancy.

This is an important survival mechanism for some species because these do not germinate unless adverse climatic conditions come to an end.

Types of dormancy

Exogenous dormancy: This type of dormancy is also known as physical dormancy, which is caused by hard seed coat, fibrous or mucilaginous (adhesives gum) during dehydration and ripening making seeds impermeable to water and gases. Thus there is prevention of some physiological processes responsible for germination. Physical dormancy can be overcome by exposing seed to high temperature and scarification which softens the hard seed coat and make it permeable for water.

Hot water treatment: It can be accomplished by soaking seeds in hot water for 5-10 minutes depending on the seed type. Exposure od seeds to hot water for loger period of time can can kill the seed.

Chemical dormancy caused by presence of some inhibitors in the seed coat which prevents the germination. i.e. fleshy fruits (citrus, grapes, tomatoes), huls of dry fruits phenols, coumarins, abscisic acid, etc. It can be overcome by acid treatment.

Acid treatment: It involves soaking of seeds in concentrated sulfuric acid for 30 to 120 minutes. Once the tratement is over, seeds are immediately rinsed with weater to drain excess of the acid otherwise delayed process may lead to considerable damage to the seeds. This is a very effective way to treat seeds with physical dormancy.

Dormancy can be removed mechanically by weakening the seed coat called as scarificatioin. Scratching or rubbing the seed coat with file or against rough surface ruptures hard seed coat which allows water penetration. This method of breaking dormancy is recommended for scarifying small quantity of seeds.

Seeds can be mechanically treated by by using following methods

- Filing (using sand)
- Chipping (using knife)
- Piercing (using needle)
- Chilling (using temperature)
- Pre-drying (using temperature)
- Pre-washing (using water)
- Pre-soaking (using warm water)

Endogenous dormancy: This type of dormancy is due to physiological and morpho-physiological factors. It can be overcome by exposing seeds to a period of moist and chilling condition. Exposing seeds to moist and chilling period is known as stratification.

Stratification: The optimum temperature for stratification ranges from 1 to 5°C, which is ideally available in most refrigerators. Physiological dormancy is satisfied by keeping the seeds in moist (substrate) soil or sand over the period of low chilling temperature. The same conditions can be obtained by keeping the seeds in a plastic bag containing a moist sand in the refrigerator for several months.

Seeds with morphological dormancy have small sized underdeveloped embryo which is shed from the plant when they are still in development stage. Such seeds require a period of moist, warm stratification for continue development of embryo. After completion of embryo development, it still has physiological dormancy which can be broken by exposing the seeds to a period of moist, chilling stratification. Seeds with morpho-physiological dormancy can take several years to germinate because they need to be exposed to summer and winter conditions.

Germination

The process of initiation of active embryo growth which results in rupturing of seed coat and emergence of a new seedling plant capable of independent existence. In other words, germination is emergence of normal sedlings from the seeds under ideal conditions of light, temperature, moisture, oxygen, nutrients *etc*. The embryo remains dormant in the dry seed, but embryo becomes active upon water imbibition and starts to grow and develop into a seedling. According to ISTA (2004) germination is defined as the "Emergence and development of the seed embryo with those essential structures which for the seed in question, indicates its ability to produce a normal seedling under favourable conditions.

Factors affecting seed germination

Several important factors affecting germination are sowing depth, water availability, temperature, sunlight and air. Besides this, certain internal conditions such as resting period, seed viability, presence of auxins and food supply to the axis are also necessary for germination to occur.

Water is a crucial imput for seed germination as it promotes various metabolic activities, respiration and activates enzymes which break reserve food material. Dry seeds contain 10-15% moisture which is very low for any vital activity. The protoplasm becomes active only when it is saturated with water. Water facilitates the necessary chemical changes in food materials & also softens the seed coat.

A suitable temperature is necessary for the germination of a seed. Protoplasm functions normally within a certain range of temperature. Higher the temperature (within certain limits), more rapid is the germination.

Oxygen is necessary for respiration of a germinating seed. The process liberates energy by breaking down the stored food and activates the protoplasm.

Light is absolutely necessary for subsequent growth. Seedlings not only elongate rapidly when raised under dark but become very weak due to lack of chlorophyll and bear only pale, undeveloped leaves.

Steps in seed germination

1. **Imbibition:** The seed absorbs water and seed coat start swelling and gets soften during imbibition process.
2. **Lag phase:** During this phase, seed activates its metabolic activities and respires more rapidly which promotes break down of reserve food material thereby leading to protein synthesis.

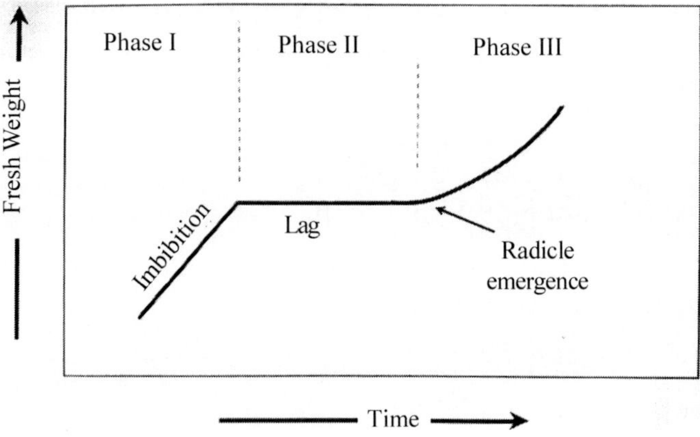

Steps in seed germination

3. **Seedling protrusion (Radicle and root emergence):** The cell division and elongation takes place which results in emergence of root and radicle out of the seed.

Types of seed germination

Based on the cotyledons or storage organ, three kinds of seed germination occur and are illustrated in the figure below:

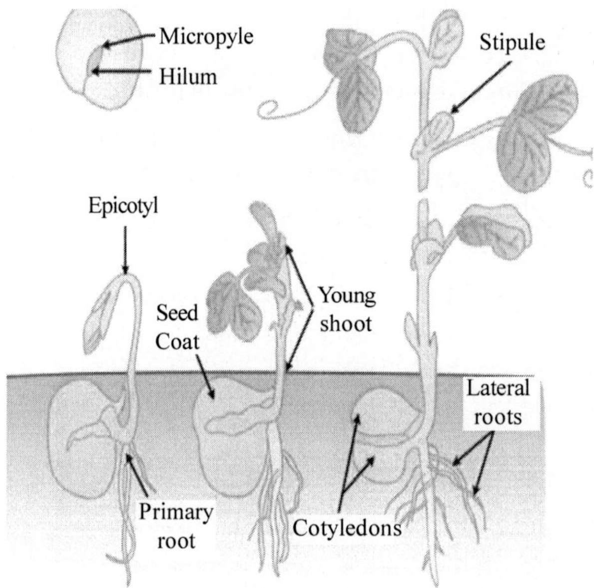

Hypogeal germination

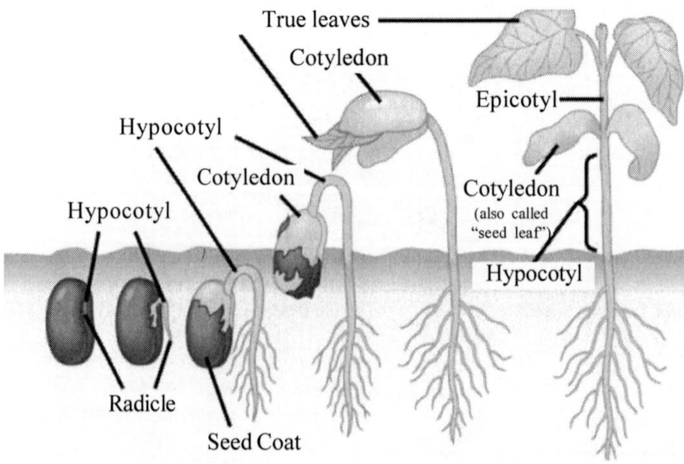

Epigeal germination

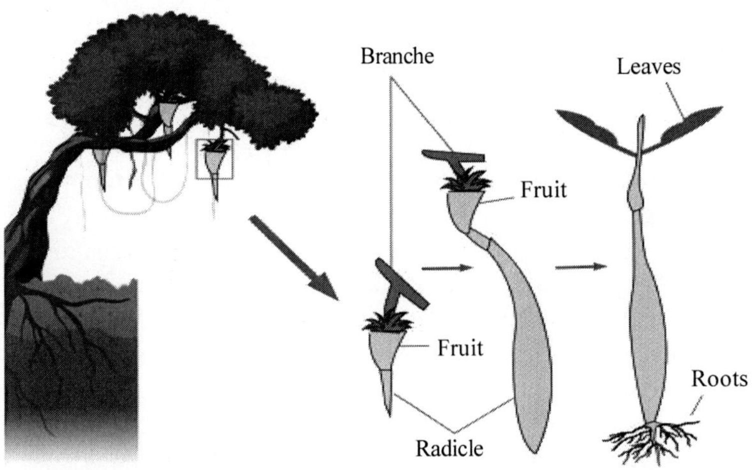

Viviparous Germination

Hypogeal germination

A type of germination in which, the cotyledons remain inside the seed coat and do not come above the soil surface. In this type of germination, radicle comes out and appears first by rupturing seed coat near the micropyle. It continues its growth and then bend into the soil. After that, the plumule grows near by the elongation of the epicotyl and starts growing upward above the soil surface. Example: Mango, peas, litchi *etc.*

Epigeal germination

A type of germination in which the cotyledons come out of the seed coat above the soil surface. In this type of germination, radicle emerge first which grows into the soil in the form of hypocotyl. Plumule growth and germination is delayed. The hypocotyl continues to grow by forming a loop and then get straighten to raise the seed above the soil surface. Then, seed splits to form two cotyledons. These cotyledons grow above the ground and extend to produce cotyledonary leaves. Examples: Papaya, Onion, Cucurbits, Tamarind, French bean, *etc.*

Viviparous Germination

In this type of germination, the embryo develops rapidly without any resting period. The embryo grows out of the seed and fruit while the fruit is still attached to its parent plant. The hypocotyl grows to a considerable length from the fruit. The radicle elongates, swells in the lowest part and gets stouter and heavier and the seedling get separated from the parent plant. Example: Chow-Chow, Groundnut, Cocoa *etc.*

Seed Enhancement: It is post-harvest treatment that improves germination and seedling growth. Various seed enhancement techniques are used to improve seed performance which increases seed vigour and germination by modifying seed emergence capabalities, ultimately resulting in improved crop yield and quality. It improves seed germination which is important in propagation and crop improvement programme. Pre-sowing treatments have been found promising and effective to get higher germination and better seedling growth in many fruit species. Seed enhancements through various methods of pelleting, coating or encrusting facilitate easy handling and planting of seeds. Seed enhancement can also deliver materials (*eg*. nutrient inoculants, beneficial micro-organisms) needed at the time of sowing. These treatments can also be designed to remove weak or dead seeds from a seed lot using upgraded, non-traditional conditioning treatments such as color sorting or x-rays. Various seed enhancement techniques have been given below:

Seed Priming: It is a pre-sowing treatment that involves the controlled hydration of seeds, sufficient to allow pre-germinative metabolic events to take place but insufficient to allow radical protrusion through the seed coat. This technique has been used in some vegetable seeds to increase the germination rate, seedling uniformity *etc*. It permits early DNA replication, increase RNA and protein synthesis, enhances embryo growth, repairs deteriorated seed parts and reduces leakage of metabolites. It is seen as a viable technology to enhance rapid and uniform emergence, high vigour and better yield.

Objectives of seed priming

- To decrease the time required for germination and emergence.
- To improve the stand uniformity, aiding in production management and increasing the chance for uniformity at harvest.
- To extend the temperature range at which a seed can germinate.
- To increase the rate of germination at any particular temperature.
- To overcome phytochrome-induced dormancy in plants such as lettuce and endive.
- Priming can reduce germination times in the field by approximately 50% upon subsequent rehydration.

Different types of seed priming are osmo priming, matrix priming, hydro priming, halo priming, thermo priming, bio priming and PGR priming.

Different types of seed priming

Osmo priming: Osmo priming is soaking of seeds prior to sowing in osmotic priming solutions [polyethylene glycol (PEG; 6,000-8,000 mol. wt.), sugar, mannitol, inorganic salts (KNO_3, KCL, $Ca(NO_3)_2$, *etc.*) and vermiculite compounds) for certain period of time. Osmo priming has a positive effect on the enhancement of seeds germination and seedlings growth, especially under stress conditions. Osmo priming helps to increase salinity tolerance in tomato and melon seeds. Osmo priming of bitter gourd seeds is a technique recommended to overcome sub-optimal environmental conditions.

Matrix-priming: Mixing of seed with water and solid material at specific proportions is known as solid matrix priming. Most commonly used solid carriers are peat moss, vermiculite, diatomaceous earth, ground lignite coal substances, charcoal, clay and sand. Seeds imbibe water slowly as a result of matrix priming thus reaching an equilibrium hydration level. Solid matrix priming improves seed vigour and germination of soybean. Solid Matrix priming of onion increases seed germination, growth and emergence under sub-optimal and optimal conditions. Sand priming increases α-amylase activity, membrane system integrity and emergence speed of corn cultivars. Solid matrix priming with calcium aluminum silicate significantly increases seedling vigour and fruit quality of okra.

Hydro priming: Soaking of seeds in simple water prior to sowing is hydro priming. Hydro priming is found to increase seed germination, seedling vigor, homogeneity of emergence, better plant stand and fruit yield. Seed priming with water is cheap and simple method, which has potential to improve seedling emergence homogeneity, germination percentage under water stress (drought) conditions and this technique can be easily used and adopted by the farmers. Hydro priming is proved to be beneficial in beans, okra, cowpea and soyabean and gladiolus.

Halo- priming: A type of seed priming in which seeds are soaked in salt solutions, which enhances germination and seedling emergence uniformly under adverse environmental conditions. Generally, calcium chloride ($CaCl_2$), potassium chloride (KCl), sodium chloride (NaCl), sodium nitrate ($NaNO_3$), manganese sulphate ($MnSO_4$), magnesium chloride ($MgCl_2$) and particularly potassium nitrate (KNO_3) salts are used for halo-priming of vegetable seeds

Bio priming: The use of beneficial micro-organisms like *Bacillus, Trichoderma, Gliocladium* in the priming medium to control disease proliferation.

PGR priming: Also known as hormonal priming. Seeds are soaked in the solution of plant growth regulators like Gibberellins, auxins, cytokinins and ethylene to enhance germination.

References

Basra, S.M.A., Zia, M.N., Mehmood, T., Afzal, I. and Khaliq, A. 2002. Comparison of different invigoration techniques in wheat (*Triticum aestivum* L.) seeds. Pakistan Journal of Arid Agriculture, 5 : 11-16.

Dutta, A.C. 2001. A Class Book of Botany. OUP India.

Ghassemi-Golezani, K., Jeddi, A.C., Nasrullazadeh, S. and Moghaddam, M. 2010. Influence of hydro-priming duration on field performance of pinto bean (*Phaseolus vulgaris* L.) cultivars *African Journal of Agricultural Research*, 5 (9): 893-897.

Heydecker, W., Higgis, J. and Gulliver, R.L. 1973. Accelerated germination by osmotic seed treatment. Nature, 246: 42-44.

Kepczynska, E., Justyna, P. G. and Kepczynski, J. 2003. Effects of matriconditioning on onion seed germination, seedling emergence and associated physical and metabolic events. Plant Growth Regulation, 41: 269-278.

Kumar, R., Misra, K.K., Misra, D.S. and Brijwal, M. 2012. Seed Germination of Fruit Crops: A Review. Hort Flora Research Spectrum, 1(3): 199-207.

Lin, J.M. and Sung, J.M. 2001. Pre-sowing treatments for improving emergence of bitter gourd seedlings under optimal and sub-optimal temperatures. Seed Science and Technology, 29: 39-50.

Mercado, M.F.O. and Fernandez, P.G. 2002. Solid matrix priming of soybean seeds. *Philippine Journal of Crop Science,* 27(2): 27-35.

Moosavi, S.S., Alaei, Y. and Khanghah, A.M. 2014. *International Journal of Agriculture and Forestry,* 4(3A): 12-17.

Ramzan A., Hafiz, I.A., Ahmed, T. and Abbasi, M. 2010. Ffect of priming with potassium nitrate and dehusking on seed germination of gladiolus (*Gladiolus alatus*). *Pakistan Journal of Botany*, 42 (1): 247-258.

Sharma, A., Rathore, S.V.S., Srinivasan, K. and Tyagi, R.K. 2014. Comparison of Various Seed Priming Methods for Seed Germination, Seedling Vigour and Fruit Yield in Okra (Abelmoschuse sculentus L., Moench). *Scientia Horticulturae*, 165: 75-81.

Singh, R. 2017. Effects of hydro priming on seed germination and vigour of *Aegle marmelos*. *Journal of Pharmacognosy and Phytochemistry*, 6(5): 446-449.

Sivritepe, N., Sivritepe, H.O. and Eris, A. 2003. The effect of NaCl priming on salt tolerance in melon seedlings grown under saline conditions. *Scientia Horticulturae*, 97(3): 229-237.

Taylor, A.G., Allen, P.S., Bennett, M.A., Bradford, K.J., Burris, J.S. and Misra, M.K. 1998. Seed enhancement. *Seed Science Research*, 8: 245-256.

Zhao, G., Zhang, T. Zheng, D. 2009. Improving the field emergence performance of super sweet corn by sand priming. *Plant Production Science*, 12(3): 359-364.

5

Nursery Management

S.K. Acharya, R.K. Jat, Mukesh Kumar and Dhara J. Barot

A nursery is a place, where seedling, saplings, trees, shrubs, and other plant materials are grown and maintained until they are placed in a permanent place.

Advantages of Nursery

1. The area being small and compact, it is convenient and easy to grow large number of seedlings per unit area.
2. Management of favourable growing conditions becomes easy and feasible.
3. Easy management of pests and diseases in compact area.
4. Establishment of disease free and virus free scion bank.
5. Ensure optimum utilization of labour, water, nutrients and other inputs.
6. Ensure easy and cheap availability of plants.
7. Promotion of export through supply and processing of quality plant materials.

Role of Nurseries in Horticulture Development

1. Production of Genetically Pure Nursery Stock

Genetic purity of planting material is essential for healthy and vigorous plant growth. Both stock and scion should be genetically pure. The planting material should satisfy all the quatitative and quantitative parameters and be easily available for further multiplication.

2. Export of Nursery Stock

Globalization has improved the chances of export of quality planting material to other countries. Special techniques and care are required for exporting the nursery material. Similarly, great care is necessary while importing nursery material from outside.

3. Employment Generation

There is a huge demand of skilled professionals for grafting, budding, potting, repotting and other nursery operations. Nursery provides employment opportunities for technical, skilled, semi-skilled, and unskilled labour. Nursery can itself be a very remunerative enterprise in the changing national and international scenario.

4. Role of Nurseries in Dry Land Horticulture

Growing drought tolerant fruit crops provide assured income to farmers. Horticultural plantations play an essential part in ecological balance thereby helping in the reduction of global warming.

TYPES OF NURSEY

A. Based on Irrigation

a) **Dry Nursery:** The dry nursery is a nursery which is maintained without irrigation or artificial watering.

b) **Wet Nursery:** The wet nursery is a nursery is a maintained by artificial water during dry periods.

B. Based on use

1. Temporary Nursery
    - This type of nursery is raised nder open especially under tree shelter or even in totally open condition.
    - In this nursery, there is no provision of permanent bed.
    - It consists of raised nursery beds.
    - Depending on the need, location of nursery area can be changed from one place to another.
2. Permanent Nursery
    A. There are generally permanently walled beds oftenly provided with overhead coverings as protection against high temperature, rain, frost *etc.*

    B. Sidewalls with drainage holes are constructed with concrete to height of 75 cm.

    C. This type of nursery may have facility of overhead irrigation system.

    D. After removal of each batch of seedlings, the soil is enriched with manures.

E. Sterilization of soil is very important component of such nursery to control soil-borne pathogens.

C. Based on location

1. Rural Nursery
   - This type of nursery is located in a village near some highway or near a railway station. The size of rural nursery is usually large because land and labour are easily available in rural areas.
   - The planting materials in such nursery are available at reasonable rates because the cost of production is less in rural areas.
2. Urban Nursery
   - This type of nursery is located in a town or a city.
   - The size of nursery is small because the land is costly and not easy available.
   - The planting material is also costly in these nurseries because the cost of production is high due to costly labour and other management practices.

D. Based on commercial basis

1. Wholesale Nursery
   - The plants are produced in large number for sale to retail outlets.
   - These nurseries are usually located in rural areas, where the land and labour are available at cheaper rate.
   - Rural nurseries can afford to expand their business without incurring much on additional expenses.
2. Retail Nursery
   - The retail seller's purchases plants from wholesale nursery and its trade is directly dependent on the demand by household users.
   - It should be located near a town or city.
   - These nurseries also deals in some important inputs like nursery tools, fertilizers, seeds, media *etc*. required for plant raising.
3. **Landscape Nursery:** The landscape nursery should be near a popular town or city because urban people require the landscape plants for beautifying their house.
4. **Mail order nursery:** It is specialized wholesale Nursery which depends primarily on a catalogue display of the stock; it offers for sale. Customers

can place order from the catalogue and get the consignment of plants requested through parcel service.

5. **Agency Nursery:** The agency nursery sales its stock through agents or sale representatives. Such Nurseries are highly specialized and usually few in numbers.

E. Based on Plants Propagated

1. **Fruit Plant Nurseries:** Fruit nurseries are specialized in production and suipply of grafts and rootstocks of fruit crops.
2. **Vegetable Nurseries:** Vegetable crops requiring nursery raising and then transplanting in main field are produced and supplied by such nurseries. In majority of instances, seedlings produced by vegetable nursery (ies) from seeds which are very expensive.
3. **Ornamental Plant Nurseries:** Ornamental crops require special practices regarding their propagation and early growth. So, these nurseries are specialzed in producing and supplying ornamental plants.
4. **Forest Plant Nursery:** A production unit that grows and supplies planting stock (seedlings and saplings) of forest trees and shrubs.
5. **Medicinal Plant Nursery:** Nursuries dealing with production and supply of planting material of plants of medicinal values are called medicinal plant nursuries.
6. **Hi-Tech Nurseries:** Tissue culture and other modern techniques are used in such nurseries for raising of seedlings in highly controlled climate.

F. Based on Ownership

1. **Public Nurseries:** Nurseries owned by some departments of the government like forest, social forestry, agriculture, horticulture *etc.*
2. **Private Nurseries:** Nursery business which is owned by an individual.
3. **Cooperative Nurseries:** Nurseries which are owned and managed by a group of individuals like self help groups *etc.*
4. **Assisted Nurseries:** Nurseries which are developed under various schemes of Horticultural Board, Employment Guarantee Schemes *etc.*

Nursery Components

1. Fence

- Prior to establishment of a nursery, a good fence with barbed wire must be erected all around the nursery to prevent tress pass of stray animals and theft.

2. Road and paths

- A proper planning for roads and paths inside the nursery will not add only the beauty, but also make the nursery operations easy and economical. This could be achieved by dividing the nursery into different blocks and various sections to facilitate various nursery operations.

3. Progeny block/Mother block

- Nursery should have a well maintained progeny or mother plant block/scion bank planted with those varieties in good demand. The rooted cuttings/ seedlings or even seeds should be obtained preferably from original breeder/ research institute from where it is released or from reputed/accredited nurseries.

4. Nursery office cum stores

- An office cum store is needed for effective management of nursery. The office building may be constructed in a place, which offers better display of all available products like fruit plants, ornamentals, seedlings, grafted plants *etc.* propagated and multiplied in that nursery with detailed information about price. A store room of suitable size is needed for storing polybags, tools and implements, packaging materials, labels, pesticides, fertilizers *etc.*

5. Sales area

- The nursery sale should easily be identified and located near to the nursery entrance. Packing facilities should be available for the customers and foe the bigger consignmemnts, receiving trucks should be directed to the loading areas by signs placing necessary signs facilitate easy and convenient loading.

6. Seed beds

- These components are essential to raise the seedling and rootstocks and should be laid out near the water source, since they require frequent watering and irrigation. Beds of a 1 m wide of any convenient length are to be made. The working area of 60 cm between the beds is necessary.

7. Nursery beds

- Raising of seedling/rootstocks in polybags require more space compared to seed beds but mortality can greatly be reduced along with high level of uniformity. Nursery beds area should also have a provision to keep the grafted plants either in trenches of 30 cm depth and 1 m wide to accommodate grafts in such beds.

8. Potting mix and potting yard

- A good potting mix is necessary for successful nursery raising. The potting mixture for different purposes can be prepared by mixing fertile red soil, well rotten FYM, leaf mould, oil cakes *etc.* in different proportions. But prior to the use of different components in potting mix, nurserymen must ensure that these components are free from soil-borne pathogens particularly nematodes. Alternatively, one can switch over to soil-less media like cocopeat, vermiculite, perlite etc. for nursery raising.

9. Compost pit

- Organic manure is an important and inevitable component of nursery. The compost pit should be constructed near the potting shed in order to facilitate its collection and ease in use.

10. Propagation area

- The propagation area is the heart of nursery operation and must be located in an area accessible to the production and potting areas. A propagation area located close to the office helps in better communication between the office staffs and the propagation managers decision making in resptec of number of specific plants to be produced. Propagation area, size and design are determined by production type, number of plants and species to be produced and marketed.

11. Media preparation and storage

- Media mixing and potting may be accomplished at one central location where potting media or media components are stored in bulk quantities. Potting media or components are stored either in loose piles or in open bins often constructed of concrete.

12. Production area

- Production or plant growing area will occupy the largest percentage of nursery land and should be adjacent to the potting area for easy movement and placement of plants in the field. Number and size of production area may vary depending upon equipment used and type of production. The layout and dimension of production areas must be known when designing the irrigation system. Production areas and roads may be modified to maximize irrigation efficiency. Most irrigation systems used in nurseries are permanent overhead delivery systems with impulse nozzles which deliver water in a circular pattern. It may be desirable to locate roads where water distribution patterns meet to ensure elimination of dry spots.

13. Packing shed

- Packaging shed should be located near the sale counter. Materials need to be packed and labelled properly prior to their delivery to the end users. Packing can be done in basket or boxes for easy handling.

14. Propagation structure

- There should be adequate provision for modern propagation structures like lath house, hot bed, cold frame, net house, polyhouse, mist chamber etc. depening upon the agro-climatic conditions of that particular area. Propagation structures are meant to provide optimum growing conditions for seed germination, rooting of cuttings and hardening of young seedling before transplanting in the field.

15. Service area

- The nursery area which offers the facilities for equipment storage and repair along with pesticides, petroleum, fertilizer facilities is termed as service area. It is usually located close to the nursery office with easy accessibility for loading and unloading of machinery.

16. Wind break

- A wind breaks consist of 2 or 3 rows of tall-medium-low height trees species planted closely together around the nursery area against the wind direction. Wind breaks reduce the wind velocity by filtering the wind with approximately 40-50% air permeability and protect the nursery area from hot and cold winds and even from severe storms.

17. Wells, sump, pipelines and generators *etc.*

- Fruit and ornamental nursery require abundant supply of water for irrigation, since they are grown in polybags or pots with limiting quantity of potting mixture.

Growing media

A growing medium described as the material used in a container, pots, polybags, plug trays *etc.* for successful raising of seedlings and saplings. Eaxmples are sphagnum moss , cocopeat, vermicompost, FYM, compost, leaf mould, saw dust, bagasse, bark, rice husk, perlite, vermiculite, soil, sand, calcined clays *etc.*

Characteristics of good media

1. The media must be sufficiently firm to provide anchorage to erminating seeds or cuttings.
2. It should be well decomposed with high C/N ratio.
3. Its volume must be fairly constant when either wet or dry.
4. It should have better water holding capacity.
5. It should be porous to drain excess water.
6. It should be free from weed seeds and harmful pathogens.
7. Slightly acidic medium is preferred.
8. It should be readily available, reusable and cheaper.

Types of media

Coco peat: It is a by-product of coconut husk which is prepared by grinding the husk. It is completely free from infestation of any pest or pathogen. It swells 15-18 times more than that of its original weight when soaked in water. It is cheaper and available in abundant quality. Coco-peat is considered best in providing aeration, drainage and life to media. It is commonly used as a medium for raising nurseries of vegetables and ornamental plants.

Perlite: It is light rock material of volcanic origin. It is essentially heat expended aluminum silicate rock. It improves aeration and drainage. Perlite is neutral in reaction and provides almost no nutrients to the media. As an inert light material, perlite is used to reduce bulk density. It does not retain moisture or hold any plant nutrient.

Vermiculite: It is heat expended mica and chemically it is hydrated magnesium-aluminum iron silicate. It is very light in weight and has minerals for enriching the mixture (magnesium and potassium) for enriching the mixture as well as good water holding capacity. It is neutral in reaction (pH) and available in grades according to sizes.

Pine bark: Bark should be milled to a maximum particle size of 5 mm prior to use sa media for seedlings production. Two key disadvantages with pine bark are the effect of its toxins on plant growth and its hydrophobic nature (difficult to wet). There have also been problems at times with purity of sample, and in uniformity of particle size following milling. Most of the toxin content can be leached by ageing in open air stacks, however bark must be made wet first. Newly stripped bark is harder to make wet following drying than the stock piled/ or weathered bark. Pine bark can be composted with the addition of small amounts for slow release fertilizers.

Hardwood sawdust: There are similar problems with sawdust to those of pine bark *i.e.* some timbers do have quite high levels of toxin. Same treatment is given to draon these toxins. Sawdust composted in the recommended manner can be combined with pine bark and with peat moss. In media composition, it is desirable to keep the percentage of sawdust (volume basis) to around 40% or less.

Soft wood sawdust: The most common softwood sawdust available is from *Pinus radiata*, which contains less toxins than some of the Australian hardwoods.

Sand: The value of adding sand as an inert mineral to any seedling mix is to obtain a balance between the other components and to give some bulk to the final product. A common belief is that sand aids in drainage component of media and sometimes, reverse may occur. For example, if sand is added to peat moss, it may fill in many of the narrow spaces between peat particles, thereby reducing loss of moisture. Also re-wetting of a dried-out mix is improved if there is a small amount of sand present. However, overmixing of peat and sand in a rotary mixer can lead to pulverising of the peat, thus changing its structural qualities. On the other hand, a large amount of sand added to a seedling mix is likely to produce dense mixtures. Likewise, there needs to be a balance between the extremes of fine and coarse sands.

Polystyrene pellets or beads: This material is used sometimes to add bulk in the media. It has no nutrient value or water holding capacity. Polystyrene is difficult to handle outside when there is a wind. Although the results are better if mixed dry with a little brown coal before moisture addition.

Brown coal: This material is known also as Lignum Peat or coal fines. The ratio of brown coal in a soil-less mix to other components needs to be low, about 20% by volume or even less. Higher volumes with adequate nutrition may cause plants to respond with lush growth and will be difficult to harden. Brown coal has a high water-holding capacity with a high level of unavailable water. Its water loss is in fact high too and it has been shown that plants transpire less water growing in brown coal than is the case with peat moss.

Peat moss: The use of peat moss for horticulture is increasing worldwide. In Australia there are a number of peat deposits, but not many are suited for seedling mixtures. A common local peat used in Victoria comes from Tasmania, but because of its trash content (old plant material), its quality does not compare with many imported peats. The fine sedge-peats are generally not suited to seedling mixtures due to their dense nature. Most peats are relatively stable against rapid decay. They are valued for their high water holding capacity; the sphagnum peats having a better balance between water and air, than most sedge peats. Even though peats have this capacity to hold water, but loss of moisture is high through transpiration and evaporation. Because of this, it is claimed that plants growing under optimum conditions would grow faster in this material.

Making a suitable nursery media

Mainly three ingredients namely cocopeat, vermiculite and perlite are used as rooting mix/medium for raising nursery. These ingredients are mixed in 3:1:1 (Volume basis) ratio before filling in plastic plug trays. It has also been critically noticed that healthy planting materials can be raised in coco peat alone, which is as good as raised in three constituents of rooting media. Coco peat from Sri Lanka and India contain several macro and micro-nutrients, including substantial quantities of potassium, sodium and chloride. So, before using cocopeat as rooting media, it is always very important to wash cocopeat with good quality water 2-3 times to remove excess of elements which are soluble in water such as potassium, sodium and chloride. Washing of cocopeat helps to reduce the electrical conductivity to a tolerance limit. Thereafter, buffering of coco peat is done with calcium nitrate @ 100g per 10 litres of water for 5 kg of cocopeat. During this process, calcium [2^+] is introduced in order to remove monovalent positive ions such as potassium [1^+] from the coconut complex. In this way, we can remove not only elements which are soluble in water but also elements which are bound to the coconut complex. Ideally, the treated water should be administered into coco-peat over a 24 hours period via a slow splinker system if possible, otherwise cocopeat can be rested in calcium nitrate solution for 24 hours. Once the resting period is over, coco-peat is rinsed with water twice and then cocopeat becomes ready to use as rooting medium for nursery production. Now-a-days, buffered cocopeat is also available in the market, which can be used directly without following the above mentioned process of washing and buffering.

Selection of a seedling tray

Plastic trays of different sizes are available in the market for nursery production, which are also popular known by the names of plug trays or pro-trays.The selection of the seedling tray depends on size of seeds, trays with 102 or 104 cavities (or plugs) having less cell volume are mostly preferred for small seeded crops like tomato, chilli, brinjal *etc.*, whereas trays with 70 or 50 cavities having more cell volume are used for large seeded crops like most of the cucurbits, papaya. Based on the size of plug cavities, plug trays with inverted cone and inverted pyramid type cavities are available in the market. It has been established that plug trays with inverted pyramid shaped cavities are very much suitable for ideal and strong growth of root and shoot system because these types of cavities offer a wider space for the development of root system. In contrary, inverted cone type cavities have limited space for proper growth and extension of roots thus cause clustering of root system.

Methods of sowing

Manual sowing

Most of the nurserymen usually follow the practice of seed sowing with hands. However, sowing of seeds with the help of a dibbler helps to maintain uniform sowing depth and speed as seed dibbler make uniform holes in the rooting medium, which is otherwise impossible while attempting sowing with hands. The dibbling depth is recommended as 3 times the seed diameter. Once the seed sowing is over, a thin layer (1-1.5 times the diameter of the seed) of either vermiculite or coco peat is spread over the surface of sown seeds. Manual sowing is a slow process and one can fill approximately 20-25 trays in an hour, hence not recommended for large scale nursery production.

Machine sowing

A seeder assembly can be used for automated tray filling, sowing, covering and watering for large scale nursery production, which make it possible to fill about 90-100 trays in an hour.

After Care

Plug Tray Placement

It is important to transfer the sown plug trays to a darkened germination room/ closed chamber maintained at a mild temperature. Small scale nurseries may not have space for a separate germination room, so one can use one of the corners for this purpose. This step is very much needed to retain sufficient moisture for germination by reducing evaporation losses. The practice of stacking many plug trays on top of each other is also followed by some nurserymen but this practice is not recommended as plug tray at the top will compress the trays beneath. Due to space constrains, germinating plug trays need to be stacked sometimes. So, it is recommended to use a ziz-zag arrangement of placing trays one over anther, which will enable trays to avoid direct contact with the media of lower one. Commercial nurseries follow the practice of putting germinating trays on nursery tables, which can be managed very efficiently by few persons.

Watering

Water quality plays very important role in deciding the health and growth of seedlings. The most common problem usually faced by most nurserymen is due to the present of dissolved salts in water. Electrical conductivity (EC) and pH are two important criteria used to adjudge the water quality, the ideal range of which are less than 1 mS cm^{-1} and 6.5-8.4, respectively. Seedling growth gets affected badly, if the water EC is high, while higher pH level is associated with higher levels of salts that will damage seedlings directly. Therefore, assessment of these quality parameters (EC and pH) can easily be done with the help of EC meter and pH meter.

A lower EC (less than 1 mS/cm) is always preferable. If the EC is too high, there are several options to reduce the problem:

- A rainwater harvesting structure may be constructed to mix rainwater with ground water to reduce the EC.
- Water softeners (such as potassium chloride) with bactericidal properties may be used to reduce water EC.
- Good drainage in the seedling trays should be ensured as it does not allow build up of salts under the high water EC at initial stage.

Suggested time and frequency of irrigation

It is best to irrigate in the morning. If seedlings are irrigated in the evening, water droplets will remain on the leaves and lead to fungal infections. If the growing medium is porous, less water is required. Watering is generally done two or three times a day depending on weather conditions. During damp weather, irrigation can be reduced to once a day. If drooping of seedlings is noticed, water immediately. Reduce watering in the last week to assist in hardening of seedlings prior to transplanting.

Nutrient supply

Fertilizers are not added to the growth medium, but nutrients are supplied to the growing seedling in the artificial medium through nutrient solution.

Basic fertilizer application

Daily application of 19:19:19 containing micronutrients (preferably Grade IV) @ 5 g per 10 litres of water, starting from the cotyledon stage till 16 days and gradually increasing the dosage every 3-4 days from 16 to 24 days, and then later stopping for hardening.

DO's and DON'Ts in nursery management

DO's

a) Adoption of double door system.

b) Prefer nuts and bolts in place of welding.

c) Work on a tarpaulin sheet while filling plug trays with media.

d) Keep watch on EC and pH of water and nutrient solution with the help of EC meter and pH meter.

e) Pack the seedling trays in plastic crates.

f) Arrange the plastic crates properly in the transport vehicle.

DON'Ts

1. Do not spray when people are working without protective clothing.
2. Do not leave gaps in the nursery structure.
3. Do not dump waste near the nursery
4. Do not allow weeds to grow at the nursery boundary.
5. Do not leave gaps while spreading weed mat that may lead to weed emergence.
6. Do not let water stagnate in and around the nursery
7. Do not keep the door open
8. Do not leave holes in the insect net.
9. Do not allow the seedlings to come into contact with soil.
10. Do not fold the seedling trays in the crate.
11. Do not stack the sown seedling trays tightly.
12. Do not allow other plants to grow inside the nursery structure.
13. Do not overwater the growing medium

Advantages of high tech nursery raising

1. Seedling can be raised under adverse climatic conditions where, it is not possible otherwise.
2. Healthy seedlings can be raised.
3. There is no chance of soil-borne fungus or virus infection to seedling as the nursery is grown in soil-less media.
4. Drastic reduction in mortality in transplanting of seedlings as compared to traditional nursery raising.
5. Early / offs-eason planting is accomplished by raising such nursery.
6. Easy transportation.
7. Weed free

8. Better root development.
9. In this technology seed rate can be reduced to the 30 -40% as compared to open field.
10. Saving the fertilizer and water of use of plug tray nursery.

Potting

- When a plant is transferred from a seed bed or a flat bed to pot, this operation is called as potting.
- Potting of nursery plants is done for making plants ready for sale such as rooted cuttings of grapes, growing plants for experimental studies like pot-culture studies, for using plants as rootstocks in certain grafting methods like grafting of mango seedlings.

References

Bhimraj Bhujbal (Ed.). 2012. Resource Book on Horticulture Nursery Management, YCMOU, NAIP, ICAR, p 264.

Dickson, A., Leaf, A.L. and Hosner, J.F. 1960. Seedling quality- soil fertility relationship of white spruce and, red and white fruits in nurseries. Forestry Chronicle, 36: 237-241.

Draft Indian Standard: Requirements for Good Agricultural Practices – India GAP part 1 crop base, 2008, Bureau of Indian Standards, New Delhi, Doc: FAD 22 (1949) C, p 28.

Report of the working group on horticulture, plantation crops and organic farming for the XI five year Plan (2007-12), Planning Commission, Govt. of India, January, 2007, p 420.

Sanjeev Kumar, Saravaiya, S.N. and Pandey, A.K. 2021. Precision Farming and Protected Cultivation: Concepts and Applications, Jaya Publishing House, Delhi.

6

Plant Propagation

Pranava Pandey and A.K. Pandey

Plant Propagation is a technique through which multiplication or production of new individuals take place from selected propagules. The methods of propagation are broadly divided in two categories:

1. Sexual method
2. Asexual method

1. Sexual method of Propagation

In this method, plants are produced from seeds and such plants are called seedlings. Seed is considered as a dormant plant which develops into a complete plant if subjected to congenial environment.

Advantages

1. This is very simple, cheap and easy method of propagation.
2. Some species of fruits, ornamental annuals and vegetables which cannot be propagated by asexual means should be propagated by this method. e.g. papaya, marigold, tomato etc.
3. Hybrid seeds can be developed by this method.
4. New varieties of crops are developed by sexual method of propagation.
5. Root stocks for budding and grafting can be raised by this method. They are hardy and develop better root system.
6. The plants propagated by this method are long lived and can resist insect-pests, diseases and drought.
7. Seed can be transported and stored for longer time for propagation.
8. Presence of polyembryony in seeds (e.g. citrus, mango) leads to the development of true- to-type seedlings as in asexual method.

Disadvantages

1. Seedlings produced by this method are not characteristically and genetically true - to - type to that of their mother plant due to segregation of characters.
2. Plants requires long juvenile period for fruiting as compared to asexually propagated plants.
3. Plants grow very vigorous, so they are not well suited for intercultural practices like spraying, pruning, harvesting *etc*.
4. It is not economical to handle larger trees, as less number of trees can be accommodated per unit area.
5. The plants which have no seeds, cannot be propagated by this method. e.g. banana, fig, Jasmine, rose *etc*.

2. Asexual/Vegetative method of Propagation

In this met, od plants are produced or multiplied by means of a vegetative part of the mother plant. This is also called cloning.

Advantages

1. The plants are genetically the same (true-to-type) as the mother plant.
2. In some fruit crops like pineapple, banana, seedless grape *etc*. which do not produce viable seeds, this is the only method of propagation.
3. This is also useful to propagate the plants that produce seed, which show difficulty in germination andhave a very short storage life.
4. The plants produced by this method have short juvenile period and come into bearing earlier than seedling plants.
5. The advantages of rootstocks can be obtained by budding or grafting susceptible varieties onto resistant/ tolerant rootstocks.
6. Plants have restricted growth, thus cultural practices and harvesting are easy.

Disadvantages

1. New varieties and hybrids can not be produced by this method.
2. Plants are not as vigorous and long-lived as the seedling trees.
3. Germplasm conservation requires lot of space and is expensive as compared to storage of seeds.

Methods of vegetative propagation

There are different vegetative propagation methods, which may be used for commercial multiplication of plants. These include:

i) Cutting

ii) Grafting

iii) Budding

iv) Layering

v) Specialized vegetative structures

i) Cutting

In this method, stem, root and leaf cuttings of the desired parent plant are used as the source of propagation. Such cuttings usually contain two or more buds for sprouting. Under favourable conditions, roots and shoots are emerged out as per polarity of the cutting and produce a new plant.

Types of cutting

Based on the plant's part used, cuttings are classified in three categories namely stem cutting, root cutting and leaf cutting. Root and leaf cuttings are generally used for propagation of ornamental plants while stem cuttings in fruit crops and some vegetable crops. Stem cutting are also classified on the basis of age and softness of stem wood as per the details given below:

a) Softwood cutting: About 3-4 months old non-lignified stem is used.

b) Semi-hardwood cutting: About 5-6 months old semi-lignified stem is used.

c) Hardwood cutting: About one and more than one year old pencil thickness stem is used. It is mainly used in propagation of fruit crops. Age and thickness of cuttings may vary from crop to crop.

Preparation of cuttings

The cuttings of a pencil size thickness and 15-20 cm length having at least three buds are made from shoots. The lower cut is given in a slanting manner just below a bud while the upper cut is given at a right angle 5-7 cm above a bud. The cuttings are usually tied in small bundles and buried in moist soil or sand for a few days for callusing of the wounds. They should not be allowed to dry. Plant growth regulators like indoleacetic acid (IAA), indolebutyric acid (IBA) and naphthalene acetic acid (NAA) encourage the formation of roots on the cuttings. The basal portions of the cuttings are soaked in a dilute solution (50-500 ppm) of any of the above growth regulators for 24 hours or basal portions of the cuttings dip in concentrated solution (500-10,000 ppm) of the growth regulators for 5-10 seconds.

ii) Grafting

Grafting is an art of joining the parts of two independent plants in such a manner that they unite together and develop into a single independent plant. Basically, it is the union of cambiums of scion (the part of graft which is to become the shoot system) and rootstock (the part which is to become the root system). Scions and rootstocks used in grafting should be compatible and similar in physiological maturity. Grafting is used primarily for woody plants, tree fruits are mostly propagated by this method. The scion may be a single bud (budding), or it may have several buds (grafting).

Important Elements for a Successful Grafting Operation

- The rootstock and scion must be capable of uniting *i.e.* compatibility.
- The vascular cambium of the scion must be placed in intimate contact with that of the rootstock.
- The grafting operation must be done at a time when the rootstock and scion are in the proper physiological stage.
- Immediately after grafting operation, all cut surfaces must be protected from desiccation.
- Proper care must be given to the grafts for a speicified period of time after grafting.

Types of grafting

a) Approach/attach method of grafting: Inarching is a attach method of grafting

b) Detach method of grafting: Veneer grafting, Side grafting, Epicotyl grafting or Stone grafting, Cleft grafting, Tongue grafting, Softwood Grafting are detach method of grafting.

1. Inarching

In this method of grafting, the rootstock is approached to the scion stick, while it is still attached to the mother plant. Alternatively the mother plants are trained to be low headed and the stock is sown under their canopy. In this method, the diameter of rootstock and scion should be approximately the same. A slice of bark along with a thin piece of wood about 4 cm long is removed from matching portions of both the stock and the scion. They are then brought together making sure that their cambium layers make contact at least on one side. These grafts are then tied firmly with polythene strip. The inarched plants are watered regularly to hasten the union. The union is complete in about 2 to 3 months. A cut is then given to the scion shoot about half way through its thickness. If the shoot does not show any sign of wilting for a week, it is completely detached from the mother plant. In case the scion shows wilting symptoms meant it requires some more days for completion of union. This method is commonly followed in mango in northern India. The best time for inarching is rainy season.

2. Veneer Grafting

In this method of grafting, a short downward 3- 4 cm long cut through cambium of the stock is given at height of 15-20 cm above ground level. At base of the cut, a second short downward and inward cut is made to join the first cut, so as to remove a piece of wood with bark (chip). An exact matching cut is also made on the base of the scion to expose cambium so that the cambial layers of both components match to each other. Then, the prepared scion is inserted into the rootstock and tied with polythene strip. After the union is complete, the stock is cut back just above the graft.

Scion sticks preparation: The scion should be prepared well before the actual grafting is done. The healthy scion shoots are selected from the last mature flush for this purpose. The selected scion shoots should have plump terminal buds. After the selection of the scion shoots, first remove the leaf blades, leaving petioles intact. After 7 to 10 days, automatically the petioles shall drop and terminal buds become swollen. At this stage the scion stick should be detached from the mother plant and grafted on the stock.

3. Side Grafting

It is similar to the veneer grafting. In this technique the scion is placed on the side of the rootstock, the cuts are made and matching done so that the scion arises at about a 30° angle from the vertical stem of the rootstock. A three sided rectangular cut about 4 x 1.25 cm is made on the rootstock at a height of about 15-20 cm from the ground level and the bark of the demarcated portion is lifted away from the rootstock. A matching cut is also made on the base of the scion to expose cambium. The prepared scion is inserted under the bark flap of the rootstock so that the exposed cambia of the two components are in close contact with each other. The bark flap of the rootstock is resorted in its original position. The graft union is then tied firmly with polythene strip. After the completion of the grafting operation, a part of the top of rootstock is removed to encourage growth of the scion. When the scion has sprouted and its leaves turned green, the root stock portion above the graft union should be cut away. Mainly used in conifers that do not root easily.

4. Cleft grafting

This is also known as wedge grafting. The cleft graft is most commonly used to top work a tree; that is, to change from one variety to another. It can be used on either young or mature trees. Branches fully exposed to sunlight and in the main stream of sap flow are more successful than those in shaded or inactive areas. This method is also useful in the nursery. The rootstock to be grafted is cut smoothly with secateurs. It is then split in the middle down to about 4 cm. The bud stick having 3 to 4 buds is trimmed like a wedge at the lower end with outer side slightly broader than the inner side. The lower bud on the scion should be located just well in to the stock making sure that the cambium layers of both the stock and scion are perfectly matched.

5. Tongue Grafting

This method gives greater surface for rootstock and scion comes into contact with each other to make a strong union. The first cut is a long sloping diagonal as much as one to two inches long. The second cut begins about 1/3 of the way down from the top of the first cut. It begins vertically, and then gradually becomes nearly parallel to the first cut surface, to create the "tongue". Identical (complementary) cuts are made in both stock and scion. Preferably the scion should be the same diameter as the stock. Stock and scion should fit together properly in such a way as to perfect interlocking. It is the predominant propagation method for apples and pear.

6. Softwood Grafting

This technique is commercially used for propagation of crops like sapota, tamarind, cashew nut *etc*. Grafting is done with mature, pre-cured scion on the newly emerging soft shoot of the rootstock which is 6-9 months old.

The basic technique involves the beheading of rootstock with a sharp knife. Then, a slit is made on the beheaded stock to insert the scion. The lower portion of scion is made in wedge shape so that both the faces of scion fit with the stock. Both stock and scion are tied with help of a polyethylene strip. The scion is then covered with a polythene bag (100 gauge) or polytubes to retain moisture of scion to keep the scion fresh till the complete union is formed. After sprouting, the bag is removed. To have better success, the leaves on the stock must be retained.

This technique is effective in dry hot weather or in areas of low precipitation where mortality of nursery raised grafts is very high.

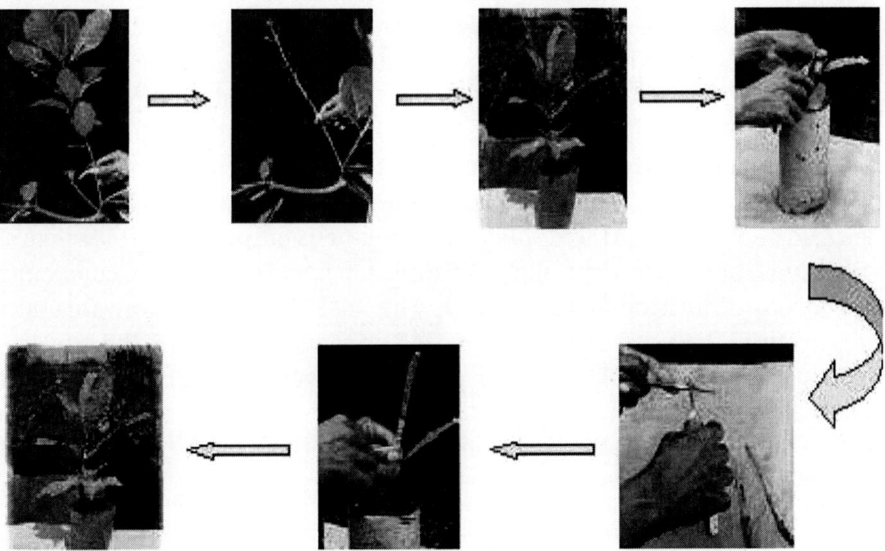

iii) Budding

Budding is also method of grafting wherein only one bud with a piece of bark, and with or without wood, is used as scion material. It is also called as bud grafting. The plant successful union of the stock and bud is also known as "budling".

Types of budding

1. T-budding/Shield budding

2. Inverted T-budding
3. Chip budding
4. Patch budding
5. I- budding

1. T-Budding/Shield budding

This method of budding is known by two names-the "T-bud" term derived from the T-like appearance of the cut in the rootstock, whereas the "shield bud" is derived from the shield-like appearance of the bud piece when it is ready for insertion in the rootstock.

T-shaped cut can be made by the vertical cut first and then the horizontal crosscut at the top of the T. As the horizontal cut is made, the knife is given a twist to open the flaps of bark for insertion of the bud. Neither the vertical nor horizontal cut should be made longer than necessary, because additional tying would be required later to close the cuts. After the proper cuts are made in the rootstock and the incision is ready to receive the bud, the shield piece or shield bud is cut out of the budstick. To remove the bark shield with the bud, an upward slicing cut is started at a point on the stem about 13 mm (1/2 inches) below the bud, continuing under the bud to about 2.5 cm (1 inches) above. The shield piece should be thin, but thick enough to have some rigidity. A second horizontal cut is then made 1.3 to 1.9 cm (1/2 to 3/4 inches) above the bud, permitting the removal of the shield piece. The next step is the insertion of the shield piece containing the bud into the incision in the rootstock. The shield is pushed downward under the two raised flaps of bark until its upper, horizontal cut matches the same cut on the stock. The shield should fit snugly in place, well covered by the two flaps of bark but with the bud itself exposed. The bud union must be wrapped with polythene tape, to hold the two components firmly together until healing is completed. When bud is sprouted, the rootstock is cutback just after at the bud graft point.

2. Inverted 'T' budding

This method is most suitable for places receiving heavy rainfall. In this method, the rootstock si given the transverse cut at the bottom rather than at the top of the vertical cut. While removing the shield piece from the budstick, the knife starts above the bud and cuts downward below it. The shield is removed by making the transverse cut 13 to 19 mm (1/2 to 3/4 inches) below the bud. The shield piece containing the bud is inserted with normal polarity into the lower part of the incision and pushed upward until the transverse cut of the shield meets that made in the rootstock.

3. Chip Budding

The species whose bark does not slip easily without tearing, may be propagated more successfully by chip budding. In this method stocks and budsticks should be 0.25 to 1 inch in diameter. The chip is removed from the rootstock by making two cuts. The first is a downward cut at a 45° angle, going about 0.25 inches through the stem. The second cut starts about 1 inch higher than the first, going downward and inward until it connects with the first cut. Matching cuts on the budstick to remove the scion, about 0.25 inches below the bud and 0.5 inch above the bud to fit the scion to the stock so that the cambium layers of stock and scion matched. Then all cut edges wrapped with plastic strips, but leave the bud uncovered. After successful union, cut back the rootstock just above the bud union.

4. Patch Budding

These methods are suitable for plants with thick bark. In this method, a rectangular patch of bark is completely removed from the rootstock and replaced with a patch of bark of the same size containing a bud of the cultivar to be propagated. Patch budding requires that the bark of both the rootstock and budstick be slipping easily. The bud patch is then transferred to rootstock and fixed smoothly at its new position and tied with polythene strip. In propagating nursery stock, the diameter of the rootstock and the budstick should preferably be about the same, about 13 to 25 mm (1/2 to 1 inches).

5. I-Budding

In I-budding, the bud patch is cut in the form of a rectangle or square, just as for patch budding. With the same parallel-bladed knife, two transverse cuts are made through the bark of the rootstock. These are joined at their centers by a single vertical cut to produce the shape of the letter I. Then the two flaps of bark can be raised to insert the bud patch beneath them. A better fit may occur if the side edges of the bud patch are slanted. In tying the I-bud, be sure that the bud patch does not buckle outward and leave a space between itself and the rootstock. I-budding is most appropriate when the bark of the rootstock is much thicker than that of the budstick.

iv) Layering

Layering is a process of development of roots on the stem while it is still attached to the parent plant. The rooted stem is become a new plant developing on its own roots under favourable condition. Such rooted stem is known as a layer.

Types of layering

1. Air layering
2. Simple layering
3. Compound layering
4. Mound layering

1. Air Layering or Marcottage or *Gootee*

In this method, layering is done on the stem in air and roots appear on it while it is still attached to the parent plant. For this one year old, pencil thickness, straight stem is selected and ring of bark about 2.5 cm width is removed from just below a bud. This portion is covered around with moist sphagnum moss that retains moisture for long duration and is wrapped with a polythene strip. Any other material, which can retain moisture for long period of time, can be used for this purpose. February–March and July-August is best time for air layering.

2. Simple layering

In this method, plants' stem bend to the ground and it's covered with soil, keeping the terminal portion of the branch uncovered. Rooting may be encourages by slight wounding in lower side of the bent stem.

3. Compound layering

In this method, stem bending to the soil as for simple layering, but alternately cover and expose sections of the stem followed provided with each section should have at least one bud exposed and one bud covered with soil. This method is most suitable for the plants producing vine-like growth.

4. Mound layering

This method is generally used for the propagation of guava and rootstocks of some temperate fruits like as apple, pear and cherry. In this method of layering, the desired plants are first established in the nursery at 1-2 m distance and the complete established plants are cutback to ground level about 2 inch above every year when they are in dormant condition. The stubs/stools are earthed up or mound with moist soil so that the new shoots coming out in spring and rainy season are covered with the soil that leads to develop roots at their basal part. These rooted shoots are then detached from the mother plant and are planted in the nursery for grafting or budding with the desired scion.

v) Propagation by Specialized Vegetative Structures

Specialized vegetative structures like as bulbs, tubers, rhizomes, corms, suckers, runners, offsets, bulbils, stolon, slips, crowns, *etc*. are also used for commercial propagation in some crops.

References

Aldriance, G.W. and Brison, F.R. 2000. Propagation of Horticultural Plants, Tata Mc Grow Hill Book Company. Inc, New York.

Bose, T.K., Mitra, S.K., Sadhu, M.K. and Das, P. 1997. Propagation of Tropical and Subtropical Horticultural Crops. 2nd Edition, Naya Prokash, Kolkata.

Edmond, J.B., Sen., T.L., Andrews, F.S. and Halfacre, R.G. 1963. Fundamentals of Horticulture, Tata McGraw Hill Publishing Co., New Delhi.

Ganner, R.J. and Choudari, S.A. 1972. Propagation of Tropical Fruit Trees, Oxford & IBH Publishing Co., New Delhi.

Garner, V.R., Bradford, F.C. and Hooker, Jr. H.D. 1957. Fundamentals of Fruit Production, McGraw Hill Book Co., New York.

Hartman, H.T. and Kester, D.E. 1976. Plant Propagation. Principles and Practices, Prentice Hall of India Pvt. Ltd. Bombay.

Mukherjee, S.K. and Majumdar, P.K. 1973. Propagation of Fruit Crops, ICAR, New Delhi.

Sadhu, M.K. 1996. Plant Propagation, New Age International Publishers, New Delhi.

Sharma, R.R. 2002. Propagation of Horticultural Crops: Principles and Practices, Kalyani Publishers, New Delhi.

7

Establishment of An Orchard

Pranava Pandey

Layout of an orchard

Layout means the way by which something is arranged or laid out. The purpose of designing a layout for an rchard is to assign the actual position for trees/ plantations, roads, water channels, buildings *etc*. The layout should be planned very carefully and implemented properly to facilitate proper care and maintenance of an orchard. The plan should provide optimum number of trees per unit area as well as sufficient space to each of the tree for proper development and ease in various cultural operations of the orchard.

Materials required for layout

Measuring tape, rope, pegs, poles and planting board.

Based on agro-climatic conditions, the following system of planting are generally followed for planting of fruit trees:

i) Square

ii) Rectangular

iii) Hexagonal

iv) Quincunx

v) Triangular

vi) Contour

i) Square system

In this system, the distance between row to row and plant to plant is same. The plants are at the right angle to each other and planted at each corner of a square whatever may be the spacing. This system provides easiness for the intercultural operations in both the directions. This is most common system followed for planting of orchards and easy to layout.

Number of plants = Area (m^2) / (plant to plant distance)^2(m)

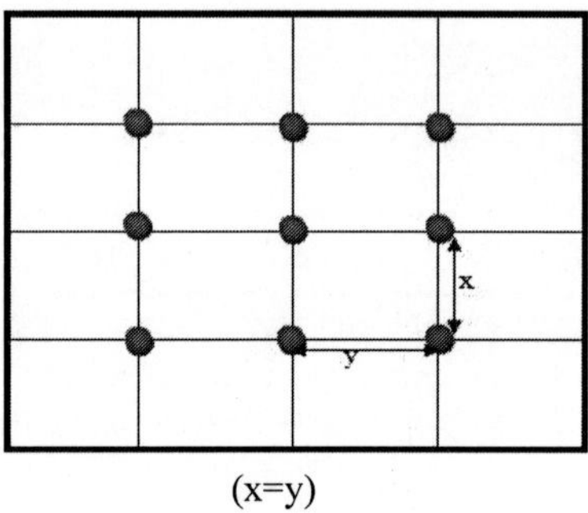

(x=y)

ii) Rectangular system

This is similar to square system, but the distance between plant to plant and row to row is not the same. In this system, row to row distance is more than the plant to plant in the row. In this system, plot is divided into rectangles and plants are planted at corners of the rectangles in straight parallel rows running at right angles. This system is also convenient to layout as well as intercultural operations in two directions.

Number of plants = Area (m^2) / Row to row distance (m) x plant to plant distance (m)

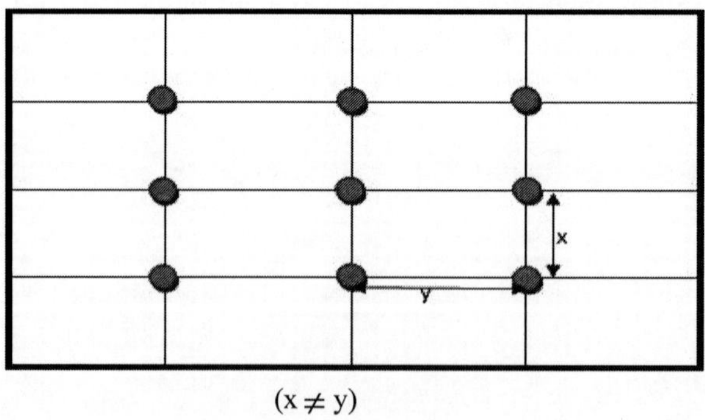

($x \neq y$)

iii) Quincunx system

It is a modified form of square system wherein one temporary, short-lived plant (filler) is planted at the centre of the square of four permanent trees that provides an additional income to the grower in the pre-bearing stage of an orchard. The filler plants may be papaya, guava, banana, fig, *etc*. When the permanent trees come into bearing stage, the filler plants are generally removed. The number of plants per acre by this system is almost doubled than the square system. This system is followed when the distance between permanent trees exceeds 8 m or more or where permanent trees are very slow in their growth and also take longer time for coming to bearing. eg. sapota, jackfruit.

Number of plants = Number of plants in square system plus (+) one row less and one plant less in each row than the square system.

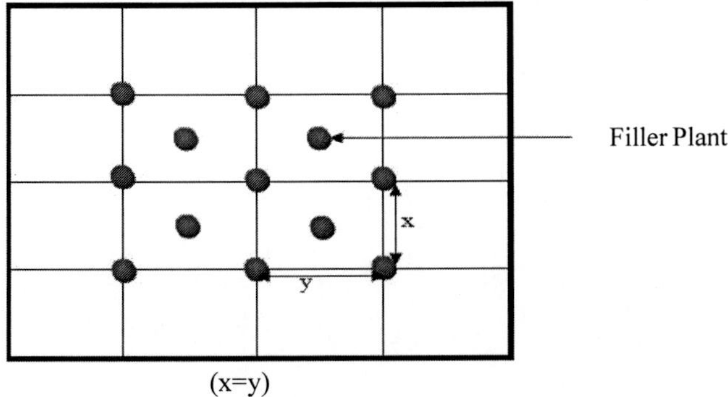

(x=y)

iv) Hexagonal system

In this system, the tree is planted at each corner of an equilateral triangle. Therefore, it is also known as equilateral triangle system. Six trees form a hexagon with seventh tree at its centre, hence termed as 'septule' system'. In this system, the distance between the rows is less than the distance between the trees in a row, but the distance from tree to tree in six directions remains the same. This system, though a little complicated for execution but can accommodate 15% more plants per unit area than the square system. This system is generally preferred where land is expensive and is very fertile with good availability of water.

Number of plants = Number of plants in a square system plus (+) 15 per cent more than the square system.

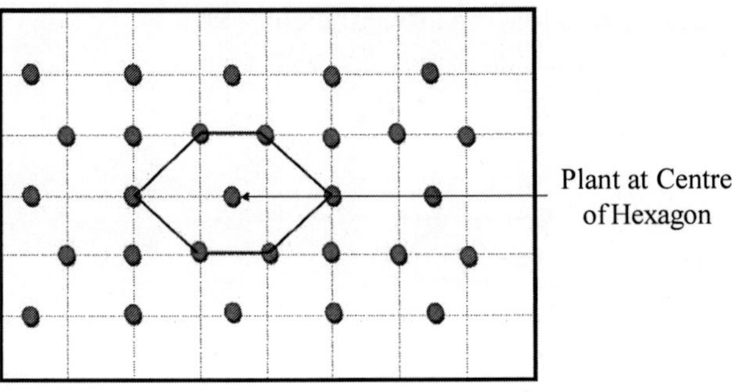

v) Triangular system

In this system, the trees are planted as similar to the square system but even numbered rows are midway between those in the odd rows instead of opposite to them. This system is based on the principles of equilateral triangle. The plants are set at fixed distances at the corners of equilateral triangles, *i.e.* the planting lines are set at sixty degrees to one another. The distance between any two adjacent trees in a row is equal to the perpendicular distance between any two adjacent rows. In this system, every even row will accommodate one plant less than in the square system while rows are equidistance.

Number of plants = Number of plants in square system minus (-) one plant less in every second row.

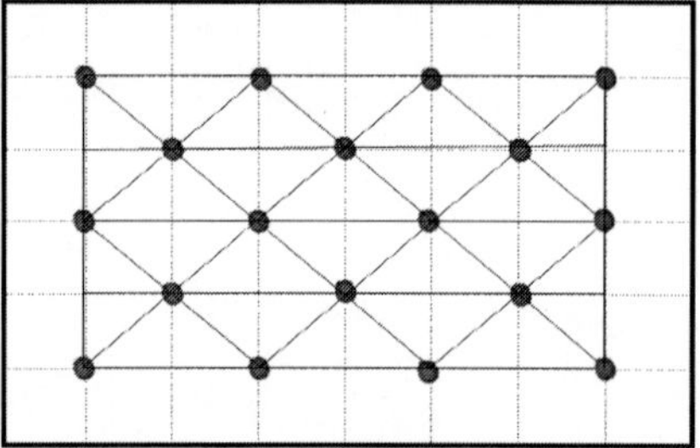

vi) Contour system

It is particularly suitable to land with undulated topography (hilly area), where soil erosion is a major problem and irrigation is difficult. By considering the major problem of land erosion and water loss, the contour line is so designed and graded in such a way that the flow of water in the irrigation channel becomes slow thus allowing more time for water to percolate in the soil without causing erosion. And in terrace system, planting is done in flat strip of land formed across a sloping side of a hill, lying level along the contours. Terraced fields rise in steps one above the other and help to bring more area into productive use and also to prevent soil erosion. When the slope is <10%, then contour bunding is practiced and if the slope is >10%, contour terracing is practiced. In this system layout is done as in square/rectangular system along the contour line.

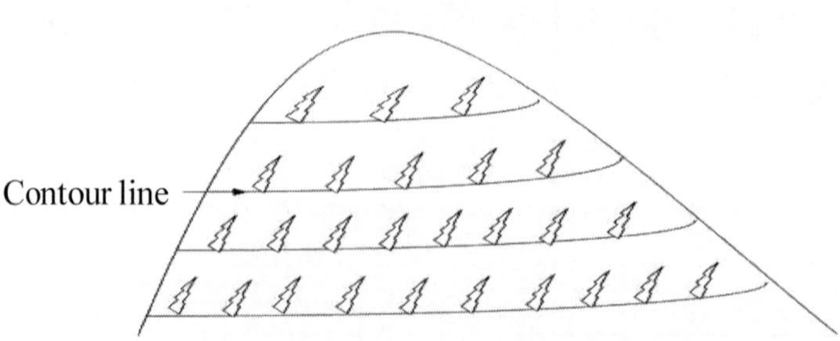

Contour line

Planting of orchard plant

Digging and filling of pits: Usually fruit plants are planted in well prepared pits, which should be dug 2 to 3 months in advance of planting. Pit size varies from crop to crop based on its root distribution pattern and soil type. While digging the pits, the top soil should be kept on one side and the bottom soil on another side separately, as the topsoil is more fertile than the bottom soil. At the time of filling the pits, the topsoil is mixed with farmyard manure or compost, leaf mould and neem cake. The pits are filled firstly with the bottom layer of soil and thereafter with the top soil mixture. Pits should be filled a few inches above the ground level for shrinkage and settlement. And time of filling of pits should be 15-30 days after digging.

Planting: At the time of planting, it is essential to check carefully the planting material about their varieties, stock-scion combination, disease and pest attack and grade of plants as preventive measures. Only healthy plants should be planted in well prepared field or pits preferably in the evening hours. The evergreen fruit plants are planted in rainy season and February-March while the deciduous fruit plants in their dormant period (December-January).

References

Bal, J.S. 2007. Fruit Growing. Kalyani Publishers, Ludhiana, India.

Bose, T.K., Mitra, S.K., Sadhu, M.K. and Das, P. 1997. Propagation of Tropical and Subtropical Horticultural Crops, 2nd Edition, Naya Prokash, Kolkata.

Edmond, J.B., Sen., T.L., Andrews, F.S. and Halfacre R.G. 1963. Fundamentals of Horticulture, Tata McGraw Hill Publishing Co., New Delhi.

Garner V.R., Bradford, F.C. and Hooker, Jr. H.D. 1957. Fundamentals of Fruit Production, McGraw Hill Book Co., New York.

Sharma, K.K. and Singh, N.P. 2011. Soil and Orchard Management, Daya Publishing House, New Delhi.

8

Training and Pruning

Pranava Pandey and A.K. Pandey

The fruit trees are grown for optimum production of quality fruits that fulfill our requirement. Therefore, proper care and management of trees is mandatory to manipulate plant growth and plant environment. Training and pruning are important operations that provide not only the proper framework to the tree but also the favourable environment leading to desired plant growth and reduction in disease and pest attack.

Training

It is a physical manipulation technique that provides proper shape, size and direction to the plant in initial stage of growth. The techniques like tying, fastening, staking, supporting over a trellis or pergola in a certain fashion or pruning of some parts are are followed.

Objectives

- To improve physical appearance and usefulness of plant by providing appropriate shape and size to the plant and maintaining proper root-shoot balance.
- To facilitate easy cultural operations including canopy management, inter-cropping, plant protection and harvesting.
- To develop strong scaffold branches with better crotch angles in all directions that are capable of bearing heavy crops over the years without breakage of limbs.
- To provide adequate sunlight and air to the centre of the tree.

System of Training in Fruit trees

1. **Central Leader (closed centre):** In this system, trunk of the plant forms a central axis that is allowed to grow freely with branches distributed all around itself, which is referred to as the central leader. Due to rapid and vigorous growth of leader stem, the tree develops a close centre and grows to long heights. Branching begins on the leader 75 to 100 cm above the soil surface. In the first year, three to four branches, collectively called a scaffold whorl are selected. The selected scaffolds should be uniformly spaced around the trunk, not directly across from or above another. Space the scaffold whorls every 50 to 60 cm up the central leader, leaving alternate areas without any branches to allow light into the centre of the tree to the desired maximum tree height. The shape of a central leader tree is like that of a Christmas tree. The lowest scaffold whorl branches will be the longest and the higher scaffold whorl branches will progressively be shorter to allow maximum light penetration into the entire tree. Example Pear
2. **Open Center (Vase shaped):** In this system the main stem is allowed to grow to a certain height and then after leader is cutback to encourage lateral scaffold branches from close to the ground providing a vase shaped plant structure that leads to a low head canopy and bulk of crop come closer to the ground. Here, sunshine is equally distributed to all the branches and trees are more fruitful and facilitate easy cultural operations (e.g. peach, apple, ber *etc.*). In the areas of high light intensity, such trees suffer from severe sun scald injuries.
3. **Modified Leader System:** This system is in between the open centre and the central leader system wherein central axis is allowed to grow unrestricted up to 4-5 years and then after the central stem is headed back and laterals are allowed. e.g. pear, apple, walnuts, *etc.*

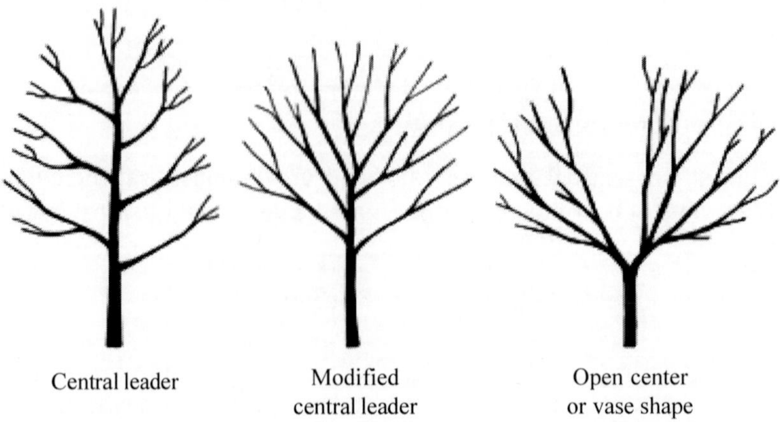

Pruning

Pruning is defined as the judicious removal of plant parts like undesirable, criss cross, diseased & dry branches, shoots, and roots with an objective to get optimum and qualitative yield.

Objectives of Pruning

- To develop the canopy in such manner that intercept maximum sunlight which leads to improves size, firmness, soluble solids and other quality attributes of the fruits.
- To distribute the fruiting wood/spur in all directions and to maintain a balance between vegetative and reproductive phases.
- To optimize the size and quality of fruit by diverting the energy into the branches of bearing fruits.
- To thin out branches to aloow more sunlight into the interior parts of the tree so that the inner parts also becomes fruitful.
- To remove diseased, damaged, insect infested and weak branches.

Methods of pruning

a) **Thinning out:** It is also known as selective pruning. It involves the selective removal of complete shoot, limb or branch from its point of origin.

b) **Heading back:** In this method of pruning, terminal portion of a branch is headed back leaving stump with buds. Generally, it is unproductive branch. And the buds left on the stump give rise to shoots which are important to the tree either being spur bearers or bearing flower buds or filling up of gaps in the tree or forming vegetative wood from which flowers may arise in the coming years.

Special Terms

Trunk: It is main stem of the plant.

Head: It is the point on the main stem from which first branch arise.

Scaffold branches: Branches arising from the head are known as scaffold branches.

Low headed tree: Trees in which scaffold branches arises under 0.7-0.9 m height from ground level are known as low headed trees. Such type of trees comes into bearing comparatively much earlier, are able to resist stormy winds more effectively and their spraying and harvesting are much easier with less expenses.

High headed tree: Trees in which scaffold branches emerges from more than 1.2 m height from ground level are known as high headed trees. Cultural management practices are very expensive in such types of trees. In the tropical climate, high headed trees are unsuitable as their exposed trunks are subjected to sunscald injury during summer.

Crotch: The angle made by scaffold limb to the trunk or the secondary branch to scaffold limb is called crotch. The crotch should be broad and the ideal crotch angle is 45°.

Leader: The main branch from ground level upto the tip dominating all other branches

Spur: Numerous shoot growths which are abundant over the fruit trees and upon which most of the fruits are borne.

Water shoots: These are extraordinary vigorous vegetative shoots which grow from the high points on the main branches in upright direction at the expense of main branches.

Suckers: These arise from adventitious buds on the roots or underground parts of the stem of the tree.

References

Bose, T.K., Mitra, S.K., Sadhu, M.K. and Das, P. 1997. Propagation of Tropical and Subtropical Horticultural Crops, 2nd Edition, Naya Prokash, Kolkata.

Edmond, J.B., Sen., T.L., Andrews, F.S. and Halfacre, R.G. 1963. Fundamentals of Horticulture, Tata McGraw Hill Publishing Co., New Delhi.

Garner, V.R., Bradford, F.C. and Hooker, Jr. H.D. 1957. Fundamentals of Fruit Production, McGraw Hill Book Co., New York.

https://ecourses.icar.gov.in

https://www.agriglance.com/

Kumar, N. 1990. Introduction to Horticulture, Rajyalakshmi Publications, Nagercoil, Tamil Nadu.

Naik, B.H. and Thippesh, D. 2014. Fundamentals of Horticulture and Production Technology of Fruit Crops, UAHS, Shimoga, India.

9

High Density Plantation

R.K. Jat, M.L. Jat, S.K. Acharya and Mukesh Kumar

The high density planting (HDP) in fruit crops is one of the new concepts of increasing the productivity without affecting the quality of fruits. The concept of high density planting was pioneered in temperate fruits in Europe at the end of 1960, since then there is rise in establishment of commercial high density orchards throughout the world. Initially HDP was followed mostly in temperate fruits *e.g.*, apple. HDP is defined as a method of planting wherein more numbers of plants are planted per unit area as compared to conventional planting density or it is a modern method of fruit cultivation involving planting of fruit trees densely, allowing small or dwarf trees with modified canopy for better light interception and distribution and ease of mechanised field operations. It is one of the important methods to achieve high productivity per unit area both in short duration and perennial fruit crops.

Now-a-days, new orchards of fruits are planted in this system with a view to produce higher fruit yield and increased profitably. In addition, there is also a need to feed the growing population and provide them nutritional security and land resources are shrinking simultaneously due to the industrialization, urbanization etc., hence it is required to develop the modern facilities to compete with the developed nation. The success of this technology in most of the fruit crops is dependent on the use of methods to control shoot growth so as to maximize light interception. The main aim of HDP is to obtain the twin prerequisites of productivity by maintaining a balance between vegetative and reproductive growth without sacrificing plant health and fruit quality.

Principle of High Density Planting

- To maximize vertical and horizontal space utilization per unit time and to harness maximum possible returns per unit of inputs and natural resources.
- To overcome low productivity.
- To reduce gestation period for early returns.

Merits of HDP

- It induces the precocity.
- Maximum utilization of land and resources
- To enhance quality production of fruit crops
- To enable mechanization in fruit crops
- Low cost per unit production
- To increase efficiency of manures, fertilizers, water, solar radiation, fungicides, weedicides, pesticides *etc.*
- Increasing pressure on land owing to diversion of farm field to various other obvious reasons as well as rising energy and land-costs, together with increased demand for fruits have made it necessary to achieve higher productivity from limited space
- High density orchards have better amenability to modern, input saving horticultural techniques such as drip irrigation

Demerits of HDP

- More competition for space, nutrients and water
- High initial establishment cost
- The close plantings between trees create considerable crowd between plant, which may result in yield decline
- The more compact and over crowded growth of canopy results in build-up of high humidity and reduced cross ventilation in the orchard, which favour more incidence of pests and diseases
- Less life span of the fruit tress
- Difficult to manage the tree canopy due to close planting
- High incidence of some diseases in HDP *e.g.* Sigatoka leaf spot & finger tip in banana
- Demand high level of technical knowledge and skill labour for the maintenance of fruit trees
- Use of dwarfing rootstocks that can control the growth of mature trees has been largely unsuccessful
- Poor availability of planting material in dwarf varieties as well as vegetatively propagated rootstocks in different horticultural crops

Components for establishing high density orchards

HDP can be achieved with the suitable use of following components:

1. Use of genetically dwarf scion cultivars
2. Use of dwarf rootstock and interstock
3. Suitable training and pruning method
4. Planting density
5. Planting geometry
6. Use of growth regulators
7. Use of incompatible Rootstocks

1. **Use of genetically dwarf scion cultivars**: It is simplest component to establish high density orchards if the trees are naturally small size. The use of genetically dwarf cultivar offers great opportunity for close plantings.

Crop	Cultivar	Desirable characteristics
Mango	Amrapali, Arka Aruna. Ratna, Dashehari	Dwarf in nature, precocious and prolific bearer
Banana	Dwarf Cavendish	Dwarf stature with high yield
Apple	Red Spur, Star Crimson, Gold Spur, Oregon Spur, Silver Spur, Red Chief	Dwarf in nature but large number of spur bearer
Papaya	Pusa Dwarf, Pusa Nanha	Dwarf stature, bearing start 25-30 cm above the ground level
Litchi	Culcuttia, China	Upright tree growth habit
Peach	Red Heaven	Dwarf and high yielder
Sapota	PKM 1PKM 3	Columnar tree shape Dwarf stature
Fig	Conardia, Excel	Dwarf stature with small canopy size
Jackfruit	PLR-1	High yielding, medium height and less spreading
Pomegranate	Amlidana	Dwarf in nature
Guava	Pant Prabhat	Less spreading and high yielder
Persimmon	Ishkei, Jiro	Dwarf in nature
Cherry	Compact Lambert, Meteor, North Star	Dwarf, self-fruitful and high yielding

2. **Use of dwarf rootstock and interstock:** High density planting is also possible due to dwarfing rootstock. Rootstocks are known to have a profound effect on the tree vigour, precocity, productivity, quality of fruits and longevity of varieties grafted on them. In this regard, standardized suitable dwarfing rootstocks for different fruit crops are given below:

Fruit crop	Dwarfing Rootstocks
Mango	Vellaikolumban, Olour,
Citrus	Alemow (*Citrus Macrophylla*), Flying Dragon (*Poncirus trifoliata var. monstrosa*), Troyer citrange
Guava	Pusa Srijan, Chinese Guava (*Psidium friedrichsthalianum*), *Psidium pumilum*
Ber	*Zizipus rotundifolia. Zizipus nummilaria*
Avacado	Colin. V-33, Maoz
Apple	M 9, M 27 (M 13 x M 9), M 4, M 7, MM 106, P 2, P 16
Plum	Pixy
Peach	Siberian C, G 677
Cherry	Colt, Charger

3. **Suitable training and pruning method:** These are essential methods to maintain shape and size of plant along with enhancement of fruit quality because overcrowding and intermingling branches are the serious problems for orchard access and adequate light interception needed for optimum photosynthesis, flowering, fruit set and quality. Pruning and production of new shoot are essential to maintain continuity of vigour and provide terminal bud for panicle emergence in mango. Pruning also helps in increasing the distribution of light through the canopy management. Various training systems adopted for maintaining the plant in dwarf shape are spindle bush, dwarf pyramid, cordon, espalier, tatura- trellis system and espalier.
4. **Planting Density:** Planting density is also an important approach for high density planting. It is determined by dwarf scion as well as rootstock and soil topography. Based on plant population, high density planting can be categorised as under:
   1. Low HDP: <250 trees per ha
   2. Moderate HDP: 250-500 trees per ha
   3. High HDP: 500-1250 tree per ha
   4. Ultra HDP: >1250 trees per ha
   5. Super HDP: 20,000 trees per ha (in apple orchards)
   6. Meadow Orchard: 70,000 trees per ha (in apple & guava)
5. **Planting geometry:** Planting system is a combination of tree arrangement and plant form. Tree arrangement in HDP system must have sufficient alleyways for the movement of farm machinery. The way trees are arranged also determines the light distribution pattern and light interception level. In fruit crops, triangular, single hedge row and double hedge row system and or square system with 4-5 m apart having enough alley space is being practiced in developed countries for HDP.

6. **Use of growth regulators:** Pruning often leads to strong re-growth of shoots in fruit crops. Plant growth regulators such as Paclobutrazol, Alar, Uniconazole, Prohexadione-calcium have been used to restrict vegetative growth and induce dwarfness. An experiment conducted at GBPUA&T Pantanagar, Paclobutrazol treatments in mango induced flowering and fruiting in new shoots produced in July after pruning without any loss in fruit quality. Application of paclobutrazol in September to November was highly effective in increasing flowering and fruiting besides reducing vegetative growth (30-35%). Thus, paclobutrazol treatments induced flowering and fruiting and helped in reducing the vegetative growth required for high density orcharding. However, uniconazole was more effective for restricting shoot growth than paclobutrazol in avocado in South Africa. Prohexadione-calcium sprays on five years old 'Hass' avocado trees reduced the growth of spring shoots compared with the growth achieved in control, but did not affect yield.
7. **Use of Incompatible Rootstocks:** Use of graft incompatible scion and stock also induce dwarfness but it was not commercially exploited. Ber cultiars when grafted onto on *Zizyphus rotundifolia* and *Z. nummularia* induce dwarfness due to graft incompatibility.

Impact of High Density Planting

Crop	Cultivar/ Rootstock	Plant Population	Impact
Mango	Amrapali	1600 Plants/ha (2.5 m x 2.5 m)	2.5 times increase in yield per hectare in Amrapali than that of the low density orchards
	Dashaheri	1333 Plants/ha (3.0 m x 2.5 m)	The average yield in high density is reportedly 9.6 tonnes compared to 0.2 tonnes in conventional planting
Citrus	Troyer Citrange	3000 Plants/ha (1.8 x1.8.m)	Increase in yield under HDP system
	Karna Khatta	1088 Plants/ha (3.0 m x 3.0 m)	
Pineapple	Kew	63758 Plants/ha	Increase in yield from 15-20 to 70-80 tonnes/ ha
Guava	Sardar	5000 Plants/ha (2.0 m x 1.0 m)	Higher yield (47.06 t/ha) under HDP system

References

Bal, J.S. 2007. Fruit Growing, Kalyani Publishers, Ludhiana, India.

Bose, T.K., Mitra, S.K. and Sanyal, D. 2001. Fruits: Tropical and Subtropical (Vol. I), Nova Udyog, Kolkatta.

Chattopadhyay, T.K. 2008. A Textbook on Pomology, Vol. 4 (Sub-tropical Fruits), Kalyani publishers, Ludhiana, India.

Goswami, A.M., Saxena, S.K. and Kurien, S. 1993. High density planting in citrus. *In*: Advances in Horticulture, Vol. 2- Fruit Crops (K. L. Chadha and O. P. Pareek, Eds.), Malhotra Publishing House, New Delhi.

http://tmnehs.gov.in

Kohne, J. and Kremer-Kohne, S. 1989. Comparison of growth regulators paclobutrazol and uniconazole on avocado. South African Avocado Growers' Association Yearbook, 12: 38-39.

Menzel, C.M. and Lagadec, M.D.L. 2014. Increasing the productivity of avocado orchards using high density plantings: A review. Scientia Horticulturae, 177: 21-36.

Mishra D.S. and Goswami, A.K. 2016. High density planting in fruit crops. HortFlora Research Spectrum, 5(3): 261-264.

Patil, M.S., Karale, A.R., Badgujar, C.D. and Adiga, J.D. 2018. Essence of Horticulture. New India Publishing Agency, Pitam Pura, New Delhi.

Ram, S. 1993. High Density planting in mango. *In*: Advances in Horticulture (Vol. 2) Fruit Crops (K.L. Chadha and O.P. Pareek, Eds.), Malhotra Publishing House, New Delhi.

Ram, S. 1996. High density orcharding in mango. Research Bulletin 122, DES, Govind Vallabh Pant University of Agriculture and Technology, Pantnagar. pp. 1-24.

Shirsath, H.K. 2013. Studies on agro-techniques in guava (*Psidium guajava* L.) cv. Sardar I. High-density planting and II. Rejuvenation of old orchard (Doctoral dissertation, Ph. D. Thesis), Mahatma Phule Krishi Vidyapeeth, Rahuri, Ahmednagar, India.

Singh, J. 2008. Basic Horticulture. Kalyani Publishers, Ludhiana, India.

10

Water Management in Horticultural Crops

Amit Kumar and Jagdeesh P. Rathore

Irrigation is one facet of the inputs which need to be carefully managed so that profits are maximized. Irrigation plays a major role in any agriculture crop production. One should be aware of different types of irrigation adapted in orchards or crop cultivation to become a successful grower. Irrigation is very important component of crop cultivation as sufficient moisture must be maintained in the soil for obtaining the optimum growth and good quality yield. The aim of irrigating a plant should be to wet the entire root zone without allowing any wastage of water beyond the root zone. The irrigation systems have to be properly devised so that the water requirements of the trees are met at the minimum expenditure without any wastage of water. Irrigation practices can be improved if a range of important factors like crop type, crop water requirements, climatic conditions, soil type, water quality, method of irrigation are taken into account.

Crop type

The type of horticultural crops grown influences irrigation practices. Annual vegetable and melon crops have high water requirement, hence irrigation management is extremely critical to their productivity during their relatively short life span (10-20 weeks). Perennial fruit crops tend to require less water and management, although critical, is not generally as critical as for annual crops. Some fruit crops (mango and cashew) require little or no water for their survival during non-flowering and non-fruiting growth periods, whereas, fruit trees (carambola, mangosteen, jackfruit and banana) from wetter tropical environments require continuous irrigation throughout the year. Crop type influences rooting depth which determines how much available soil water the plant is able to tap into. Effective root depths vary from 15-30 cm for vegetable crops to 80-100 cm for many fruit crops (mango, citrus).

Crop water requirements

The crop water requirement can be calculated using indirect evaporation based models, or directly measured using soil moisture monitoring devices. In theory both methods should be used, the former method allows the grower and irrigation designer to know how much water is likely to be required during the cropping season and how best to deliver the water so that the crop peak demand for water can be met. The latter method allows the grower to monitor water use during crop growth and to make any small adjustments which are required.

Soil type

Soil type influences irrigation management due to its ability to store varying quantities of water depending on their texture. Sandy soils hold the least and clays hold the most of water. Most soil profiles are made up of various texture classes, hence the water storage capacity depends on the cumulative storage capacities of the various layers within the profile.

Water quality

Plants require good quality water for oveall growth and development. In areas where water is being extracted from limestone aquifers the lime (bicarbonate) levels may present some management difficulties, particularly if using overhead irrigation where the leaves are wetted. Most serious water quality related problems are as a result of high salt (sodium chloride) levels in the water.

Irrigation method

Method of application of irrigation is an important part of the management process. The three broad categories, over-head (high pressure), micro-sprinkler (medium pressure) and drip (low pressure) systems have to be managed in their own specific ways. Technically, drip irrigation systems are the most efficient in terms of units of water delivered per units of water pumped. Micro-sprinklers are again more efficient than over-head spray irrigation. The type of system utilised will depend on: water availability, crop type, soil type and the preferred option. It is of vital importance that the system is hydraulically correct *i.e.* it is able to deliver the same amount of water at the required emitter operating pressure to all areas of the crop or orchard. The system must also be designed to deliver sufficient water for the crop peak requirements, as under- watering can be an extremely costly exercise.

Timing of Irrigation

The timing of irrigation can be a crucial part of managing an orchard. Some tree species, e.g. mango responds to a dry period by flowering earlier and more profusely than would otherwise occur if they were continually irrigated. The flowering response is also probably influenced by temperature and the effect of irrigation management may not be as evident in a year during a cool dry season. Other species which may respond positively to a dry period are: cashew, rambutan and durian but the response and the length of dry period required will depend on the species.

Irrigation methods

There are many irrigation methods available to adapt based on soil, crop type, water availability and method of planting.

A. Surface method of irrigation

This method is also known as traditional irrigation method. This is an old and still the most common irrigation methods used in most of the countries since centuries. Surface irrigation is classified into four types according to the supply of water.

1. Flooding method
2. Borderstrip or Bed method
3. Basin method
4. Furrow method

Flooding method of irrigation

This type of irrigation method consists of opening water channel in a field so that the water is allowed freely in all directions of land and covers the whole surface of the land. This method is the most inefficient method of irrigation as only about 20 to 25% of the water is actually used by plants or crops. The rest of the water would be lost as a runoff, seepage and evaporation. In this method, water distribution is uneven which is visible from uneven crop growth. This method of irrigation is not suitable for agriculture crops that are sensitive to water logging conditions.

Advantages of flooding irrigation

1. This method is useful with irregular topography and the land requiring extensive levelling.
2. This method is useful for high density crops.
3. Useful where plenty of water is available at cheaper and affordable rates.
4. Useful where soil erosion is less.
5. Set up cost and operation costs are less as compared to other methods.

Disadvantages of flooding irrigation

1. There is a huge loss of water in this method.
2. This will cause excessive soil erosion on steep lands or fields.
3. This method induces loss of manures and fertilizers.
4. This method encourages weed growth.

Borderstrip or Bed method of irrigation

In this method, the field is levelled and divided into small beds/strips surrounded by bunds of 20 cm to 30 cm high. Little irrigation channels should be provided between two adjoining rows of beds. For high value plants or crops, the beds might be still smaller especially where water is not abundant and moreover costly. This procedure can be adopted in many soil types except sandy soils and is suitable for high value plants. It requires levelled soil or field for uniform water distribution. This method provides a more effective way of uniform application of water in the field. This method is suitable for plants in lines or sown by broadcast method.

Advantages of Borderstrip or Bed method of irrigation

1. More area can be brought under cultivation with a lesser expenditure.
2. This method of irrigation requires less labour.
3. This method of irrigation causes lesser erosion when compared to flood irrigation.

Disadvantages of Borderstrip or Bed method of irrigation

1. This method of irrigation is not ideal for all types of plants/crops.
2. In this method, uniform or balanced supply of water is tough.
3. This method of irrigation is not ideal for all soil textures.
4. This method also requires fairly abundant water supply.
5. Land must be levelled for uniform water distribution.
6. The initial cost of this method is high as it demands land levelling.
7. Water drainage should be provided in this method.

Basin method of irrigation

This irrigation method is much more suited to horticultural crops. In this method, a raised platform known as '*thanvla*' (basin) is shaped around trees/plants and plants are connected to one another through drains and water reaches from one tree to another. This process is not suitable for cereals crops. This method of irrigation is most suitable when the size of this plot to be irrigated is small. The basin could be square, rectangular or circular form based on planting method. Ring and basin are widely used for irrigating fruit trees. It is also efficient in using water however its initial cost is high.

Advantages of basin irrigation

1. With this method, varying level of water supply is possible.
2. This method saves time and rapid irrigation is possible.
3. No loss of water by run-off.
4. There won't be any manure/fertilizer loss because very minimum soil erosion is expected.
5. This method is economically viable due to less investment.
6. More trees get benefited with this irrigation method.

Disadvantages of basin irrigation

1. This method of irrigation is not suitable for all crops mainly useful for horticultural crops.
2. Some wastage of water is observed in this method.
3. This method of irrigation may encourage spreading of diseases among trees.
4. If the land is not properly levelled, the initial cost is expected to go high.
5. This irrigation process is not suitable for soils that disperse easily and readily form a crust.

Furrow method of irrigation

This system is practiced in newly planted orchards. Taking the plants in the centre, an irrigation channel of 20 cm depth and 60 cm width is prepared. Water is allowed to flow in these furrows from the main channel. There is water saving in this system as only limited area is wetted. In orchards, two furrows on each side of the rows are generally made. Furrow system is suited to such lands, which have a moderate slope to the extent of 1 to 2% if the water is to run freely and reach the ends of the furrows. Where the slope is sharp, the furrows are made to follow the contour more or less closely.

Advantages of furrow irrigation

1. This method provides a facility to irrigate large areas at a time.
2. This irrigation method can be adapted in most of the soil textures.
3. Very easy to use in any crop with row planting method.
4. This irrigation system is easy to install and maintain.
5. This method saves labour since once the furrow is filled, it is not necessary to give water a second time.
6. This method is a comparatively cheaper method when compared to other mentioned methods.
7. This method provides enough quantity of water to the plants.

Disadvantages of furrow irrigation

1. Due to the imbalance in the flow of water, wastage of water is possible with this method.
2. Furrow irrigation is not ideal for all types of crops.
3. Skilled labour is required to manage this type of system.
4. Water drainage must be provided.
5. Excess of water penetration at the head than at the end, which may result in variation in vigour and growth of trees.

B. Sub-surface method of irrigation

Subsurface irrigation or sub irrigation may be natural or artificial. Natural subsurface irrigation is possible in which an impervious layer exists under the root zone. Water is permitted into a series of ditches dug upto the impervious coating which then moves laterally and also wets root zone. In artificial subsurface irrigation, perforated or porous pipes are laid out underground below the root zone and water is directed into the pipes with suitable means. This method of irrigation is very rarely found in India and used in other parts of the world.

Advantages of sub-surface irrigation

1. It is very efficient in using water as evaporation loss can be reduced almost completely.
2. Maintenance cost is not very high.

Disadvantages of sub-surface irrigation

1. Initial investment cost is high.
2. There is a risk of soil getting saline or alkaline and neighbouring soil ruined due to heavy seepage.
3. Applicable only on levelled land for successful subsoil irrigation.

C. Drip or trickle method of irrigation

This is a system of irrigation which supplies water to the plant equivalent to its consumptive use. This is a highly water use efficient system of irrigation having very less irrigation water requirement. Especially in arid region, drip irrigation system is very beneficial technique of irrigation. In arid areas, there are two basic constraints of surface irrigation, first one is the indulated land terrain and the second one is limited availability of water. Turning the land level is very costly venture and also with low water availability, getting production becomes a question. As the water is applied through drippers, the system naturally takes care of limited water availability. A drip system has four basic components: suction, regulation, control and discharge which are accomplished by water lifting pump, hydrocyclone filter, sand filter, fertilizer mixing tank, screen filter, pressure regulator, watermeter, main line, laterals and drippers. Water lifting pump is essential to supply water with pressure. After lifting, the water passes through hydrocyclone filter, sand filter, fertilizer mixing tank and screen filter and ultimately through drippers. Through hydrocyclone filter, relatively coaser particles, through sand filter relatively finer particles and through screen filter very fine particles are filtered. These filters are essential for smooth running of water through laterals and drippers, otherwise choking of laterals may take place. There are two types of drip systems.

High pressure drip system: This system works at operating pressure of 30 psi or more.

Low pressure drip system: This system operates at less than 30 psi pressure.

Advantages of drip irrigation

1. Water saving to the tune of 30-70%.
2. Increases yield and fruit quality.
3. Higher returns per unit area and time.
4. It saves labour cost.
5. Improves water penetration.
6. Eliminates soil erosion.

7. Reduces weed growth.
8. Saving in fertilizers and chemicals (40-60%).
9. Better pest and diseased management.
10. Eco-friendly technology.

Disadvantages of drip irrigation

1. The initial cost of commercial drip irrigation is high.
2. High maintenance requirement
3. Salinity hazard
4. High technical know-how is required.

D. Sprinkler method of irrigation

In this method of irrigation, water is spread into air and allowed to fall on the ground surface somewhat resembling the natural rain. Water is forced under pressure through small nozzle/orifice which gets broken up into droplets and fall back on the ground. Slow circular revolution is impacted to the nozzle for uniform coverage of the ground surface. The rate of application should not be more than the infiltration rate of the soil. Sprinkler irrigation can be adopted where land levelling is uneconomical and other method of surface irrigation cannot be carried out. It is designed to minimize cost towards labours, fertilizer and irrigation.

Advantages of sprinkler irrigation

1. Sprinkler system provides uniform distribution of water across the crop.
2. It can be used in almost crops expect paddy & jute.
3. System can be adopted under varied topographic conditions and especially suitable to steep-slope and irregular topography.
4. It can eliminate surface runoff of irrigation water (runoff elimination)
5. Protects the crop against frost and high temperature.
6. Reduces labour cost for irrigation as compared to surface method.
7. Saves energy and cost involved in channel preparation.
8. Saves fertilizer and water.
9. Gives higher water use efficiency.

Disadvantages of sprinkler irrigation

1. Not suitable for very fine texture soil (<4mm)
2. More evaporation losses.
3. Require clean, water free from debris, sand slit and clay particles.
4. Saline water can not be used.
5. Initial cost is high.
6. High operating power is required (5-10 kg/cm)
7. Sprinkling may sometimes be encouraging for the spread of disease.
8. Ripening softy fruits need protection from the spray.

E. Pitcher or pot method of irrigation

This system is especially a boon in arid regions in view of limited water supply. A pitcher filled with water is buried in the periphery of individual tree where feeding roots are confined. The water is released slowly and slowly with the micro-pores available in the wall of a pitcher and takes care of irrigation requirement of the plant. The numbers of the pitchers per plant depend upon the spread of the tree. Normally for a plant having 3 m spread all around, 4 to 5 pitchers per plant would be sufficient.

Advantages of pitcher or pot irrigation

1. In this method, only the place near the pot gets irrigated not the entire place.
2. Water seepage below the ground is also in minimum quantity.
3. It is the best method for horticultural crops.
4. Once the pitchers are fixed, irrigation could be done for six years, which reduces expenditure.
5. This does not require high technical knowledge.

Disadvantages of pitcher or pot irrigation

1. Irrigation in this technique is possible in a limited area.
2. This technique requires clean water because contaminated water could cause blocking of minor holes of a pitcher.

F. Bubbler method of irrigation

This system has the goodness over drip irrigation system in respect of energy saving, avoidance of infiltration and getting rid of clogging of drippers. In bubbler system, water is carried through inexpensive, thin walled, corrugated plastic pipe of appropriate diameter suitable to carry water at low pressure. From the surface ditch, water passes through pipe and the risers (vertical tube of 1-3 cm diameter) provided in the underground buried pipe and ultimately reach to the plant. The height of risers is adjusted to control the discharge of a particular volume of water as per requirement of the plant.

G. Funnel method of irrigation

This system is an outcome of innovation by a farmer of Rajasthan. In this system, a funnel or a bottle of about one litre capacity is fixed or is tied along the plant. The water is applied in the funnel or bottle regularly. As the bottle or funnel is fixed at the side of plant, the applied water reaches directly to the feeding roots and the plant survives with the use of very little water. This system is very useful for newly planted crops whose water requirement is less.

References

Goyal, M.R. 2015. Water and Fertigation Management in Micro Irrigation. Apple Academic Press, United States.

Goyal, M.R. and Singh, A. 2017. Micro Irrigation Engineering for Horticultural Crops: Policy Options, Scheduling, and Design. Apple Academic Press, United States.

Majumdar, D.K. 2013. Irrigation Water Management. PHI Learning, India.

Mazumdar, B.C. 2004. Orchard Irrigation and Soil Management Practices. Daya Publishing House, India.

Quinn, N., Tsakiris, G. and Gupta, M. 2020. Agricultural Water Management: Theories and Practices, Elsevier Science, United States.

Tei, F., Nicola, S., Benincasa, P. (Eds.). 2017. Advances in Research on Fertilization Management of Vegetable Crops. Springer International Publishing, Germany.

11

Fertilizer Application in Horticultural Crops

Pranava Pandey

Fertilizer is a substance of natural or synthetic origin, applied to soils or to plant tissues (usually leaves) for the supply of one or more nutrients essential to the plant. It is classified in two types, inorganic fertilizers and organic fertilizers. Both the fertilizers play major role to the crops if they are properly and appropriately applied. To obtain maximum benefits, it is important to apply fertilizers at an appropriate time, amount and in a precise manner. The method of fertilizer application should be selected appropriately taking into consideration the age and stage of crop plants, their root extensibility, type of orchard management, kind and amount of fertilizers to be applied. The fertilizers should be given over the area of crop plants, where their active roots are spread. In fruit trees, fertilizer should be given in restricted area *i.e.* in the surrounding area of about 1 to 1.5 m away from the trunk of the trees in active root zone.

When to apply the fertilizers?

Majority of the crops require nutrients at different stages like growth & development, flowering and fruit setting time. In fruit crops, stages like new flushes emergence, floral buds differentiation and fruit development require more nutrients. Time of application of fertilizers may also vary depending on types of fertilizers, need of the crops, soil and climate.

Methods of fertilizers application

The various methods of fertilizers application are as follows:

i) Broadcasting

In this method, broadcasting of dry granular fertilizer in a uniform pattern over a large area/field is done. It is generally followed for lawn applications and in full bearing or closely bearing orchards. Follow broadcasting when leaves are dry so that concentrated fertilizers do not stick to the leaf surface, otherwise may cause burning spots.

ii) Top dressing

Fertilizer can also be applied to specific plants or plant rows using the top-dressing or side-dressing method. This is also useful in those fields which show deficiencies of major or micro nutrients in specific pockets.

iii) Band\ Strip placement

Fertilizers placement can be made either in bands or trenches around the trees or fertilizers are drilled or injected into the soil. This method is commonly used when fertilizers are required in small quantities at young atage of a plant and when P and K fertilizers need to be incorporated into the root zone or in case of plants with poorly developed root systems.

iv) Foliar application

Soluble fertilizers can be applied as a spray or drench over the foliage where it will be absorbed through the green tissues of plants. Generally, plants respond to foliar applications very fast when the nutrient levels are relatively low. Foliar application results in fast absorption of nutrients for quick plant response. Nitrogen, phosphorous, potassium and a number of trace elements are readily absorbed through the leaves. Therefore, foliar feeding practices are effective as an emergency treatment. Moreover, foliar sprays include micronutrients application or nutrient which cannot easily be supplied through soils. Poorly performing annual crops or sometimes perennial are often treated with a foliar application required solution of fertilizer (s). Foliar applications are most effective during early growth. The use of surfactants or spreaders, such as soap solutions, can enhance nutrient uptake through leaf tissue. To avoid leaf burn, nutrient sprays should not be applied during hot weather. While applying pesticides and nutrients, it is advisable to check the comapatibility between them. Application of concentrated fertilizers should not be done too close to the trunks. Each fertilizer application should be followed by copious irrigation.

v) Fertigation

The practice of supplying crops with fertilizers *via* the irrigation water is called fertigation or Fertigation is a process of application of water soluble fertilizers or liquid fertilizers through irrigation water. The nutrients are applied via fertigation directly into the wetted volume of soil immediately below the emitter where root activity is highly concentrated. In fertigation, plants receive small amounts of fertilizer early in the crop season, when plants are in vegetative stage. The dosage is increased as fruit load and nutrients demand grow and then decreased as plants approach to the end of its crop cycle. This gives plants the needed amount of fertilizers throughout the growth cycle rather than a few large doses. Timing, amounts and concentration of fertilizers applied are easily controlled in fertigation. Fertigation allows the plants to absorb up to 90% of the applied nutrients, while granular or dry fertilizer application typically shows absorption rates of 20 to 60% (Table 22.1). Fertigation ensures saving in fertilizer (40-60%) due to better fertilizer use efficiency and reduction in leaching. Fertilizer use efficiency (%) using different methods is tabulated below:

Fertilizer use efficiency (%) in fertigation

Nutrient	Soil application	Drip + soil application	Drip + fertigation
N	30-50	65	95
P_2O_5	20	30	45
K_2O	60	60	80

Fertigation scheduling

Factors that affect fertigation module are soil type, available NPK status, organic carbon, soil pH, soil moisture at field capacity, available water capacity range, aggregate size distribution, crop type and its physiological growth stages, discharge variation and uniformity coefficient of drip irrigation system. Fertilizers can be injected into the irrigation system at various frequencies once a day or once every two days or even once or twice a week. The frequency depends upon system design, irrigation scheduling, soil type, nutrients requirement of a crop and the farmers' preference.

The efficient fertigation schedule needs following considerations

1. Crop and site-specific nutrient management.
2. Timing of nutrient delivery to meet crop needs.
3. Controlling irrigation to minimize leaching of soluble nutrients below the effective root zone.

Methods of fertilizer injection

Fertilizers can be injected into drip irrigation system by selecting appropriate equipments.

Commonly used fertigation equipments are

- Venturi Injector
- Fertilizer Tank (By-pass system)
- Injector Pump
- Automatic Fertigation Controller

Steps for programming fertigation

- Calculate peak water requirement of a crop.
- Calculate fertilizer dose to be injected according to fertility status and targeted yields.
- Select suitable fertilizers according to solubility, compatibility, nutrient ratio.
- Make a stock solution (1:10), dissolve fertilizers completely in plastic container.
- Check the suction rate of fertigation equipment selected. Do the calibration if required.
- Feed data to computer / controller, if automatic fertigation controller is being used.
- Stop all leakages from micro-irrigation system.
- Operate micro-irrigation system till field capacity in soil is achieved. Follow irrigation schedule.
- Adopt pulse injection fertilizers.
- Adjust pH and EC by adding Nitric acid, Sulphuric acid *etc.* if required.
- Adjust the concentration of fertilizers in effective root zone by adjusting volume / flow of water.
- Check the pH, EC of fertilizer solution coming from last dripper.
- Avoid over irrigation.

Points to ponder while adopting fertigation

- For precise placement of both water and fertilizer, it is necessary to use pressure compensating drippers or inline drippers instead of micro tubes.
- Daily feeding is most desirable, if not possible feed on alternate day.
- Urea can be combined with WSF for meeting the N requirements.
- Fertigation should be done at the end of irrigation period, run the drip system for 5-6 minutes after the completion of fertigation.

- It is not advisable to use a very concentrated stock solution (generally not more than 10%).
- The fertilizer solution should be compatible with the quality of water into which it is being injected.
- Do not inject fertilizers in combination with pesticides or chlorine.
- The injection point must be upstream of the filter system so that the filter will remove any un-dissolved fertilizers or precipitates that occur.
- Select fertilizer solutions to help adjust water pH if necessary.
- The time needed to distribute the fertilizer should be less than the time needed to supply enough water to the field.
- Do not over irrigate because this will lead to leaching of some of the nutrients out of the root zone.

Care to be taken during fertigation

- Install micro-irrigation system in greenhouse as per design.
- Stop all leakages from micro-irrigation system.
- Maintain field capacity all the time, do not over irrigate.
- Check the suction rates of fertigation equipments.
- Calculate the correct fertilizer dose to be injected.
- Concentration of fertilizer in effective root zone should be as per the growth stages of crop.
- Use pressure compensating drippers.
- For correct fertigation, fertilizers should be injected in last 40-50 minutes of total irrigation then continue drip irrigation for 5-10 minutes after completion of fertigation.
- Fertilizer doses should be as per soil analysis or targeted yields.

Forbidden mixtures

- Calcium nitrate with any phosphates or sulphates
- Magnesium sulphate with di or mono ammonium phosphate
- Phosphoric acid with iron, zinc, copper and magnesium sulphates

Recommended combinations of water soluble fertilizers

Materials used to supply nutrients for greenhouse crop production are chosen based on several factors including cost per unit of nutrients, solubility in water, ability to supply multiple nutrients, freedom from contaminants and ease of handling. The most commonly used fertilizer materials for greenhouse crops are tabulated below:

Sources of nutrients and their solubility

Fertilizer Grade	N : P : K	Solubility (g/l) at 25 °C
Urea	46:00:00	1100
Ammonium nitrate	34:00:00	1920
Ammonium sulphate	21:00:00	750
Calcium nitrate	16:00:00	1290
Magnesium nitrate	11:00:00	710
Urea ammonium nitrate	32:00:00	1183
Potassium nitrate	13:00:46	133
Mono ammonium phosphate (MAP)	12:61:00	375
Potassium chloride	00:00:60	340
Potassium sulphate	00:00:50	110
Potassium thiosulphate	00:00:25	1550
Mono potassium phosphate (MKP)	00:52:34	230
Phosphoric acid	00:52:00	1906
NPK	19:19:19 or20:20:20	300-400

Use of two or more fertilizer tanks separates fertilizers that interact and cause precipitation. Placing in one tank the Ca, Mg and micro-nutrients, and in the other tank, the phosphorus and sulphate compounds enables safe and efficient fertigation.

Advantages

- Nutrients and water are supplied near to the active root zone leading to a greater absorption by the plants.
- Fertilizer use efficiency through this method ranges between 80-90%, which helps to save a minimum of 25% of the nutrients supplied.
- As water and fertilizer are supplied evenly to all the crops through fertigation there is possibility of getting 25-50% higher yield.
- Saves rrigation water, fertilizer, time, labour cost and energy use.
- This is not only reduces the production costs but also lessens the potential of groundwater pollution caused by the fertilizer leaching.
- Suitable for compound and ready-mix nutrient solutions containing small concentrations of micronutrients which are otherwise very difficult to apply accurately to the soil.

References

Goyal, M.R. 2015. Water and Fertigation Management in Micro Irrigation. Apple Academic Press, United States.

Hu, C. 2019. Fruit Crops: Diagnosis and Management of Nutrient Constraints. Elsevier Science, Netherlands.

Kolay, A.K. 2007. Manures and Fertilizers. Atlantic Publishers & Distributors Pvt. Ltd., India.

MAFF, 2000. Fertilizer Recommendations for Agricultural and Horticultural Crops, 7th Edition, The Stationery Office, London.

Rao, K.M. 2005. Textbook of Horticulture. Macmillan Publishers India Limited, India.

Sanjeev Kumar, S.N. Saravaiya, S.N. and Pandey, A.K. 2021. Precision Farming and Protected Cultivation: Concepts and Applications. Jaya Publishing House, Delhi.

Tei, F., Nicola, S., Benincasa, P. (Eds.), 2017. Advances in Research on Fertilization Management of Vegetable Crops. Springer International Publishing, Germany.

12

Juvenility and Flower Bud Differentiation

Pranava Pandey

Juvenility

Juvenility is referred to the phase of plant growth following germination from seed during which flowering does not occur and the bud meristem is not competent to respond to seasonal environmental inductive cues, and hence plant remains vegetative. The changes from juvenile to mature characteristics are well known as phase changes phenomenon. Plants are unable to initiate flowering after germination and have to undergo in a juvenile developmental phase. This transition period (phase) is a gradual and continuous process until the flowering takes place. The length of this phase varies considerably between different species. In annuals like Arabidopsis, the juvenile phase is typically very short, whereas in perennials like hybrid aspen, it is several years.

A plant is considered juvenile from the time it is at seedling stage until it is mature and capable of initiating flowering. Length of juvenile period in most of woody plants is still considered as a big problem. Indeed, it affects the plant breeding efficiency, plant propagation, and selection of new cultivars. In fact, juvenility is a complex phenomenon still not fully understood. Juvenility period extends from seed germination until the plant attains the ability to flower. Furthermore, all the stages namely seed germination, juvenile growth period and transition to the reproductive stage until fruitful inflorescence development are long and not always synchronous in the seedling population. It arrives at the age 30-40 years in *Fagus sylvatica* L., and at 10 years or even more in olives. Juvenile period is affected by both environmental and genetic factors. Attempts made in past to overcome juvenile period in most woody plants on the environmental level have shown very limited success.

Flower bud differentiation

The mechanism by which a shoot apical meristem changes its anatomy to generate flowers or inflorescence. Anatomical changes begin at the edge of the meristem, generating first the outer whorls of the flower-the calyx and the corolla, and later the androecium and gynoecium.

Differentiation is the process of specialization in terms of shape and function.

The floral differentiation of fruit trees involves the transformation of leaf buds on branches into a floral bud. The change from the vegetative state to the reproductive state is one of the most dramatic events in the ontogeny of a plant. Flowering leads to an exciting succession of events like anthesis, fruit set, fruit development, maturation and ripening. It provides opportunity for the propagation of the species and assists in crop improvement through genetic recombination.

In general, fruit crops pass through a juvenile stage during which flowering does not occur naturally. Once flowering capacity is attained, the plant may respond to environmental cues such as light (especially relative lengths of light and dark periods), temperature and nutrition.

Flower bud formation requires a series of changes in the differentiation pattern of apical or axillary buds. Flower bud development is a highly complex process that is characterised by two distinct physiological phases:

a) Bud initiation and

b) Floral bud development.

Flowering factors

Flowering factors that regulate the timing of flowering in plants are

a) External and

b) Internal factors

a) External factors

Flowering is influenced by various external factors like photoperiod, light quality, vernalization and other relavant environmental factors like ambient temperature besides nutrient status and moisture stress.

b) Internal factors

Three major theories attempt to explain the internal factors or chemical control of the transition to flowering.

1. Florigen concept
2. Nutrient diversion hypothesis
3. The model of multi-factorial control

1. Florigen concept

In the 1930s, Chailakhyan postulated the existence of universal flowering hormone 'florigen' a hypothetical substrate. He proposed that florigen consists of two components, gibberellin and anthesin, both of which are presumed to be required for flowering. Florigen is produced in the leaves and acts in the shoot apical meristem of buds and growing tips. Anthesin is again a hypothetical substance. According to this hypothesis, gibberellins are required for flowering in LDPs (long day plants) and anthesin for flowering in SDPs (short day plants).

2. Nutrient diversion hypothesis

This hypothesis postulates that induction, whatever the nature of the involved factors, is a means of modifying the source/ sink relationships within the plant in such a way that the shoot apex receives a higher concentration of assimilates than under non inductive conditions. The concentration of sugar should be greater than that of nitrogenous compounds for flowering to take place. In mango except for cv. Baramasi it has been found that higher starch reserve, total carbohydrates and C/N ratio in the shoots favour flower initiation.

3. The model of multi-factorial control

Floral evocation and morphogenesis can be achieved by the application of various chemicals including carbohydrates, PGRs and PGR antagonists.

References

Bose, T.K. 1985. Fruits of India: Tropical and Subtropical. Naya Prokash, India.

Early, M., Bamford, K. and Adams, C. 2013. Principles of Horticulture. Elsevier Science, United Kingdom.

Rao, K.M. 2005. Textbook of Horticulture. Macmillan Publishers India Limited, India.

13

Unfruitfulness

Pranava Pandey

The world "Unfruitfulness" means a state or character of being unfruitful or barrenness. It is a vital constraint in so many fruit crops leading to a great loss to the farming community. In contrary, 'Fruitfulness' is the state where a fruit plant is not only having the ability of flowering and fruiting, but also carry on these fruits up to maturity. The inability of the plant to do so is known as 'unfruitfulness' or 'barrenness'. Despite of profuse flowering, poor fruit retention in an orchard has been found owing to poor fruit setting at initial stage and subsequently abscission of the fruitlets. All the trees of an orchard do not bear flowers and fruits in the same extent or regularly and sometimes fail to bear flowers and fruits under similar conditions while other trees bears heavily. Such type of characteristics of trees may be attributed to unfruitfulness. Thus, all the problems affecting fruit retention upto final stage can be considered under unfruitfulness of a tree. Therefore, it is a very complex problem of an orchard and needs proper understanding for economical and effective management to maintain sustainable production of fruit crops.

CAUSES OF UNFRUITFULNESS

There are innumerable causes perceived behind unfruitfulness. Majorly, It may be due to poor root shoot ratio, lack of balance between vegetative and reproductive (fruiting) growth, heavy crop load which leads to restriction of fruit bud production and finally short of the crop in the following years. It can also be due to the unfavourable environmental conditions that lead to improper flowering and poor fruit set. Impotency, sterility, incompatibility and abortion of embryos are also some of the causes of the unfruitfulness. In broader sense, the causes of unfruitfulness can be divided into two categories:

i) External factors

ii) Internal factors

i) External factors

1. Environmental factors
    - Temperature
    - Rainfall
    - Wind
    - Frost
    - Hailstorm
    - Cloudy weather
    - Light intensity
2. Disturbed water relations
3. Nutrient supply
4. Rootstocks
5. Pruning
6. Age and vigour of the plant
7. Locality/region
8. Seasonal influence
9. Spraying of chemicals during flowering
10. Insect-pest and diseases
11. Miscellaneous factors

ii) Internal factors

a) Evolutionary tendency

b) Genetic Influence

c) Physiological factors

Description of unfruitfulness' factors

i) External factors

1. Environmental factors

Temperature

Temperature is directly related to physiology of a plant. Each crop and species require a definite and optimal range of temperature for growth, development, flowering, pollination and fruit set. These processes of the plants are affected directly or indirectly when temperatures deviate from its optimum range. Examples-

- In case of litchi crop, sex of flowers is decided well before they open. If day temperatures of 26 °C and above during a period of 5 to 7 weeks before flowering promotes the induction of male flowers while female flowering is favoured by temperatures of 19 °C and below during the same period.
- Influence of temperature on fruit set in papaya has also been reported such as sex reversing male plants emerges due to temperature fluctuations.
- The pollens of most of the temperate fruit crops like apple, pear, peach, plum, cherry, walnut, pecan nut *etc.*, germinate at temperature 10 °C or above, but the fertilization process is inhibited strongly if the temperature falls below 4.4 °C
- Higher temperature and dry atmosphere appear to be associated with the increased production of pollens per anther in mango crop
- Temperature indirectly influence fruit set through its effect on the pollinizer varieties and the activity of pollinators.

Rainfall

Rainfall is an important natural phenomenon of prevailing weather condition which helps in proper growth and development of a plant. But at the time of blossoming, it is considered as one of the most important limiting factors affecting fruit set and ultimately the fruitfulness in fruit crops. Rain created favourable microclimatic conditions to insect-pests and diseases. It washes off the pollens from the anthers as well as affect stigmatic receptivity of the flowers.

Wind

Wind is an important pollinator in fruit crops *viz.*, walnut, pecan nut, hickory, hazelnut, coconut *etc.* This type of pollination is known as Anemophilous.

- Speedy wind adversely affects insect pollination because pollen carrying insects work more efficiently in a calm wind.
- Wind also causes the stigmatic fluid to dry leading to reduced stickiness of pollens and poor germination.

Frost

Frost at the time of flowering may either completely destroys the blossoms or destroy the sexual organs, thereby influencing the fruit set and ultimately the fruitfulness of the crops *e.g.* mango, banana, guava, litchi, *etc.* In temperate crops, spring frost severely affects fruitfulness. In certain climatic conditions, it may be responsible for a regular bearing cultivar or a plant to become an irregular bearer.

Hail storm

Hail storm might be able to create a significant harmful effect to the crops if it occurs at any time between flowering and fruit developmental stages. It destroys all the flower buds and causes injuries in almost all the developing fruits. In hail-prone areas, anti-hail nets should be stretched over the orchards to reduce the losses.

Cloudy weather

It is an important limiting factor for making conditions conducive to development and spread of insect pests and diseases, which leads to unfruitfulness in many plants. For example, powdery mildew, a most serious disease of many fruit crops, usually appears immediately just after cloudy weather which leads to drying up and drop. If cloudy weather occurs during peak flowering period, powdery mildew may become most destructive to the crops.

Light intensity

Light intensity is an impotent factor to determine flowering and fruiting in almost all the plants especially the deciduous fruit plants. Effect of light intensity has been reported in high altitude apple orchards that produce apples with intense colour than lower altitude and in strawberry, as the development of pistils and stamens take place only when plants are exposed to certain light intensity. Poor fruit set, yield and colour development in high density orchards have been mainly attributed to dense and intermingle tree canopy, rendering poor light penetration, for which canopy management is recommended.

2. Disturbed water relations

For proper production and yield, adequate supply of water is required. Disturbance in water supply in soil-

- May lead to the formation of an abscission layer, resulting in the dropping of the blossoms, leaves, fruits *etc.*, thereby affecting fruitfulness of the plant and
- May be responsible for disturbance in C:N ratio and composition of other chemicals, which are responsible for fruit drop.

It has been reported that excess moisture causes flower and fruit drop in so many fruit crops. In general, it is suggested to avoid irrigation of orchards during flowering as it may sometime lead to complete shedding of flowers as reported in citrus and guava.

3. Nutrient supply

Nutrient supply in balanced form is always required for proper growth and yield. Excess supply of nutrients not only causes vigorous vegetative growth of the plants, but it adversely affects flowering and fruiting. Balanced supply of nutrients maintain desirable C:N ratio which is a major prerequisite for flowering and fruiting in fruit crops. Imbalance supply of nutrients disturbed C:N ratio of plants and leads unfruitfulness.

Examples

- Aonla crop in fertile soil does not perform well as in nutrient deficient soil. And Jonathan apple when grown on fertile land in Victoria (Australia) behaves as self-unfruitful but becomes self-fruitful when grown on land with poor fertility.
- The Hope grape (perfect flower variety) and Muscadine group produce hermaphrodite flowers only when proper nutrient supply is given. While under poor nutrient status and faulty management practices, these varieties produce most of the flowers staminate ones.
- Some cultivars of strawberry produce perfect flowers and are productive when grown under ordinary management cultural practices, but produce only little pollen for satisfactory crop when grown on rich soils.

4. Rootstocks

Rootstocks are providing root system to the scion parts through which plants absorb water and nutrients. Rootstock and scion relationship have much significance in the commercial fruit production. Growth patterns and behaviours of one plant that are produced by combining of two different plants or genotypes by budding or grafting is different from those that would have occurred if each component part had been grown separately. The various characters induced in scion cultivars due to rootstocks may be desirable or sometimes undesirable. In several way rootstocks exerts ignificant influence on flowering and fruit set of scion cultivars.

Example: Seedling plants are relatively late bearing and susceptible to abiotic stresses and diseases but this problem can be overcome by grafting or budding with desirable rootstocks. It has been reported in fruit crops like Troyer Citrange (Citrus) for quick decline tolerant, Dogridge (Grapes) for salt tolerant, Pusa Srijan (Guava) for dwarfing character and Khirni (Sapota) for strong root system *etc*. In this way rootstocks play an important role in the fruitfulness of the plants.

5. Pruning

Pruning is an intercultural operation adopted in many fruit plants where undesirable and diseased parts of plant are judiciously removed to maintain the rhythm of fruiting especially in deciduous fruit plants. However, it can also be done in evergreen fruit plants depandeing upon the requirememnts.

Example: Heavy pruning is required for good quality yield in peach and Cane-pruned grapevines produce flowers and fruits better than the severe spur-pruned grapes.

6. Age and vigour of the plant

Age and vigour of the plant directly affect yield and fruitfulness of crops. This factor is directly associated to canopy structure of the plant. If canopy of fruit plant is not properly managed, it turns to poor flowering and fruit setting due to imbalance supply of photosynthates. In general, after long run every plant becomes shy in bearing and lead to unfruitfulness.

Example

- Young and vigorous Aonla trees are known to show profuse flowering but fruit set is not in proportionate due to presence of large number of male flowers.
- Young and vigorous plum trees are known to produce higher proportion of defective pistils than older trees of the same variety.
- Young vigorous apple trees often fail to set fruits under controlled cross-pollination, whereas old and less vigorous trees of the same variety set freely.

7. Locality/region

It has been seen that fruit retention and fruit setting on trees of similar genotype is often much better in one location than in another. In similar condition, a tree flowers and set fruits profusely in one side but get no fruits on another side. It is just because of environmental factors of a particular location which influences directly the fruiting.

Example

- An alternate bearing character in mango is commonly found in cultivars of Northern parts (Dashehari, Langra, Chaunsa) of India.
- The South Indian variety (Neelum and Baneshan) of mango don't perform better under North Indian conditions.

- The American grape varieties have been found to be self-sterile in Columbia while it is perfectly self fertile in California, which is in the South.
- The Jonathan apple is almost sterile in Victoria (Australia), whereas it is fertile in USA.

8. Seasonal influence

Seasonal influence is directly associated with environmental factors which aries from one location o another. So, it is typical to differentiate the influence of locality from that of season, because seasonal variations at a particular place is about to be similar as locality variations in case of environmental factors.

Example

- Change in sex form of plum variety from protandrous to protogynous with a change in seasons.
- Sex reversal in papaya is associated with environmental factors.
- Change of sex with season (Japanese Persimmon, Grapes Hybrid – Vitis riparia x Vitis Labrusca)

9. Chemical Spraying during flowering

Chemical (insecticides and fungicides) spraying on fruit trees is generally menat to prevent infestation/infection by insect-pests and diseases, but spray during flowering time may hamper the movement of pollinators thereby affecting fruit setting and finally fruitfulness of an orchard.

Example

Jhumka, the disorder of mango crop is due to excessive spraying of insecticides, which result in decline in pollination and pollinating insects. Mercury, lead, and arsenicals cause major injury to the pollen and pollen carrying insects, if they are sprayed on trees during flowering.

10. Insect-pest and diseases

Insect-pest and diseases directly attack the tender parts of a plant and the flowers are amongst them and so attack by numerous insect-pests, fungal, viral and bacterial diseases on flowers under congenial condition render them unfruitful too. But in other way, insects are useful for pollination and fruit setting in fruit crops also.

Example

- Houseflies strictly do pollination in mango, but in favourable condition mango hopper may infest its blossoms in such a number as to blast off all the flowers of a tree or even in the whole orchard.
- Among diseases, blight is most serious disease, which limits fruit setting in pears considerably and others like Scab (Apple and Pear), Powdery Mildew (Mango, Grape, Ber *etc.*)

ii) Internal factors

Internal factors do not express itself easily rather observing their outcomes. It is also divided in three categories as:

a) Evolutionary tendency

b) Genetic Influence

c) Physiological factors

a) Evolutionary tendency

Evolutionary tendency of a genotype or species naturally promote variability and maintain their vigour via adapting cross pollination and fertilization. Such type of species faces difficulty in self-fertilization resulting in enhanced unfruitfulness. The factors involved are:

i) Imperfect/defective flowers

a) Monoecious: "One House" - both sex parts on same plant but at different sites Exp. Cashewnut, Annona, Aonla, Ber, Litchi, Jackfruit *etc.*,

Hermaphrodite: Male and female parts within the same flower.

b) Dioecious: Plant is either a female or a male. Male and female sex parts present on separate plants Exp. Papaya, Date palm etc.

ii) Heterostyly

Difference in the length of style.

a) Pin type flowers- Presence of long style and short filaments for example, almond and carambola.

b) Thrum type flowers- The presence of short style with long filaments, for example sapota and pomegranate.

iii) Dichogamy

Variation in maturity time of sex organs.

a) Protandrous- Ber, Passion fruit

b) Protogynous- Annona, Mango, Avocado

iv) Stigmatic receptivity

Stigma receptivity is the ability of stigma to support pollen germination, which is a crucial stage in successful fertilization. It limits the effective pollination period (number of days during which the pollination is effective in producing a fruit and is determined by the longevity of ovules minus the time lag between pollination and fertilization) in fruit crops. Cessation of stigmatic receptivity has been associated with degeneration of stigma and rupture of papillar integrity in kiwi fruits.

v) Abortive flowers or aborted pistils or ovules

This is found in the developing flowers, pistils and stigmas. Various causes of flower abortion in different fruit crops are hereunder:

Example

Peach- Degeneration of nucleus, embryo abortion, Grapes- Degeneration of nucleus, Mandarin- Abnormal pistil, Litchi- Embryo abortion, Olive- Pistil abortion, Kiwi fruits- Pollen degeneration, Apple- Defective embryo, defective ovules.

vi) Impotency of pollens

In this case, prevention of fruit setting may be due to various reasons like failure of pollination, pollen sterility or even nutritional deficiency. Examples:

- Unfruitfulness in Muscardine grape is due to defective pollen.
- In Ber, Illaichi variety is male sterile.
- In Citrus, Kagzi Lime and Eureka Lime are male sterile due to non-viable pollens.
- Pollen sterility is common in many Olive cultivars.

Stout and his co-workers (1916) recognized that such type of unfruitfulness is mainly due to the following three internal factors:

i) Sterility from impotence

ii) Sterility from incompatibility and

iii) Sterility from embryo abortion

i) Impotence

- Sterility from impotence arises when one or both the sex organs fail to develop the fruit properly.
- The impotence may be complete, in which either no flower or no sex organs are formed, or it may be partial, in which either stamens or pistils are abortive.

ii) Incompatibility

- Sterility from incompatibility arises, when although the sex organs are completely formed, but they fail to function properly.
- The pollen grains are unable to germinate freely on stigma or stigma is not compatible with the pollens.

Thus due to incompatibility, the properly developed gametes fail to unite together, although the sex organs are completely functional.

iii) Abortion

- Even after the proper pollination and fertilization, the abortion of the embryo takes place before their maturity.

b) Genetic influences

i) Incompatibility

Under self-pollination condition, there could exist an inability of functional male (pollen) and female (ovule) gametes to set fruits or seeds of same variety (self-incompatibility) or of the different variety (cross-incompatibility). Self incompatibility is more common in fruit crops like apple, pear, sweet cherry, almond, avocado, fig, mango, citrus, olive, etc. than cross incompatibility like apple, pear, sweet cheery, European plums and almond.

- Sporophytic self-incompatibility: In mango, self-incompatibility is reported in cvs. Dashehari, Chausa and Langra.
- Gametophytic self-incompatibility: In loquat varieties like Golden Yellow, Pale Yellow pollen tube penetrate into stylar canal up to $1/4^{th}$ to $1/3^{rd}$ of its length but do not go further below even after 72 hours of pollination.

ii) Sterility due to hybridity

In general, the wider the cross, higher is the chance of degree of sterility.

- 'Troth Early' peach x 'Wild Goose' plum, - 'Mule' bears flowers abundantly, but the flowers neither have stamens nor pistils.

- Peach x Sand cherry - Kamdesa
- Pear x Quince - Pyronia
- Sterility appears in number of hybrids between *Vitis rotundifolia* and *Euvitis* group due to hybridity.

iii) Inter-fruitfulness and inter-fertility

The ability of two plants or two varieties to set fruits and develop seeds with each other's pollen is called as inter-fruitfulness or inter-fertility.

Example: Aonla, Smyrna fig, Datepalm, some varieties of Grapes.

iv) Reciprocal crossings

In some fruit crops, it has been reported that if a certain crossing is proved to be sterile, its reciprocal crosses rea also sterile and if one variety is proved to be incompatible with the other, those two are likewise incompatible with each other. However, in others, a certain set of crossing has been fruitful, but the reciprocal crossing being the sterile.

Example

- Tragedy plum (European type) is a good pollinizer for several varieties of Japanese type, but fail to set fruits when Japanese plum varieties is used for the Tragedy plum.
- *Vitis Vinifera*, *V. labrusca* and *V. cordifolia* species of grape set fruits freely either with *V. rotundifolia* and *V. munsoniana*, but when *V. rotundifolia* and *V. munsoniana* are used as pollen parents either for *V. Vinifera*, *V. labrusca* or *V. cordifolia*, they never set fruits freely.

c) Physiological influences

i) Slow growth of the pollen tube *eg*. Clementine mandarin

ii) Poor pollen germination

iii) Premature or delayed pollination *eg*. Kagzi Kalan

iv) Nutritive condition of plant: Poor nutritive condition of a plant not only affects vegetative growth, but also results in the production of defective pistils and poor pollen production.

v) Crop load: This factor is associated with alternate bearing in which fruit crop has exhaustion of food reserves due to heavy crop load in 'on' year, leaving very little or no food reserves for flowering and fruiting in the next 'off' year.

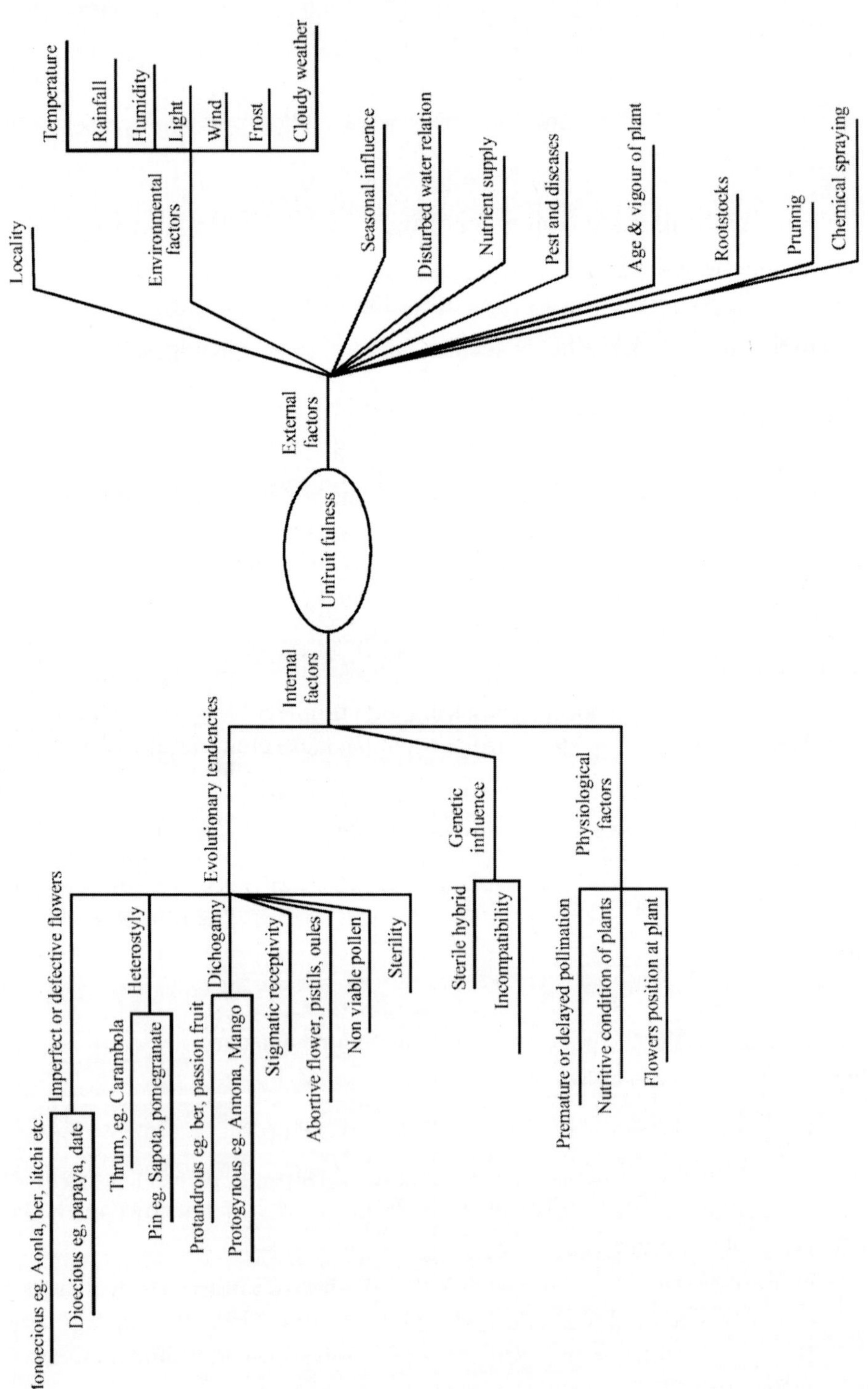
Unfruit fulness
External factors
Locality
Environmental factors
Temperature
Rainfall
Humidity
Light
Wind
Frost
Cloudy weather
Seasonal influence
Disturbed water relation
Nutrient supply
Pest and diseases
Age & vigour of plant
Rootstocks
Prunnig
Chemical spraying
Internal factors
Evolutionary tendencies
Imperfect or defective flowers
Monoecious eg. Aonla, ber, litchi etc.
Dioecious eg. papaya, date
Heterostyly
Thrum, eg. Carambola
Pin eg. Sapota, pomegranate
Dichogamy
Protandrous eg. ber, passion fruit
Protogynous eg. Annona, Mango
Stigmatic receptivity
Abortive flower, pistils, oules
Non viable pollen
Sterility
Genetic influence
Sterile hybrid
Incompatibility
Physiological factors
Premature or delayed pollination
Nutritive condition of plants
Flowers position at plant

REMEDIAL MEASURES

Maintaining proper ratio between vegetative and fruiting growth of trees

For maintenance of proper ratio of plant growth, different techniques like pruning and training, use of dwarfing rootstocks, growth retardants, compact & short-internodes scions are utilized which provide proper canopy with wide crotch angled branches. Here, pruning is most important thing because it is not only maintaining canopy only but the loss of carbohydrates reserve also through pruned wood.

Control of pollination

Pollinizers and pollinators both play major role in pollination of fruit crops which is a most essential activity for fruit setting. So, population of the both in an orchard should be maintained at proper level on each part. Pollinizer varieties are most important for temperate fruit crops and major pollinator bee population can be maintained by placing bee-hives that gives great help in ensuring satisfactory fruit set. Therefore, arrangement of bee-hives and inter-planting of suitable pollinizer cultivars must be followed at the time of orchard layout.

Preventing frost damage

This is mainly concerned with temperate fruit crops. During dormancy, risk of frost is minor but in the spring when the tree starts to break dormancy, the major damage occurs. Unfortunately, there is no cure for frost damage; a tree affected during its spring growth and bloom will have to wait until the next year to fruit. The exact temperature that causes bud kill will vary by the type of fruit, variety, bud maturity, and length of cold exposure. Most fruit trees in bloom can withstand temperatures as low as - 2.2°C for 30 minutes, with only 10% bud kill (and thus no reduction in harvest).

The first step to prevent frost damage is to select a variety of tree that is cold hardy. If region experiences typically a late spring frost, select a variety known to have a late bloom time. If temperatures are expected to drop too low once the tree begins budding or blooming, or if snow is predicted, it's time to take action. Small trees can be wrappped in frost blanket bags for the duration of cold snap. And For larger trees, spraying with frost shield provides protection to young leaves and flowers without damaging them or preventing pollination from occurring. Frost shield as aclobutrazol and Ethephon's repeated spray may delay bud break and blossoming in pear and prunes, respectively.

Proper nutrition

Proper nutrition is always desirable for realizing optimum yield in crop plants. Generally, it is advised that application of fertilizers a few days before the emergence of blossoms is generally believed to favour flowering and fruit set.

Application of plant growth regulators

The unfruitfulness problem of fruit crops may be due to poor fruit set and abscission of fruits at various developmental stages, which can be minimized by the application of plant growth regulators. Some of the findings regarding application of plant growth regulator to overcome unfruitfulness in fruits are: Litchi- TIBA/KNO_3: increase pollen fertility, Apple- Cultar (pactobutrazol): Increases yield, Apple - GA_3+NAA at petal fall stage: Increases fruit set and initiation, Apple - GA_{4+7} at any time between one and 40 days after blooming: Reduces June drop of fruitlets.

Use of suitable rootstocks

Rootstocks serve as the root system of the tree thereby maintaining nutritional balance. It also plays very important role in fruitfulness by controlling tree vigour, providing tolerance to salinity and drought, resistance to diseases and insect pests *etc*. Thus, suitable rootstock increases fruitfulness of trees. Screening of mango rootstocks to salinity has shown that the polyembryonic cultivars 'Olour'and 'Bappakai' could withstand higher level of salinity. Apple rootstocks namely: M9, M7, M4 and M1 have ability to induce 50% or more bloom in the 5th year in 'Starking delicious' apple by controlling tree size resulting in higher yield efficiency.

Conclusion

Thus, unfruitfulness is a general and complex problem found in all fruit crops and their cultivars at certain level due to various internal and external factors. But it can be managed to some extent by proper planning and layout of orchard from initial levels of establishment. Adaption of necessary corrective measures like selection of fruit crops and cultivars on the basis of soil and climatic factors, maintenance of varietal diversity, rejuvenation of old orchards, thinning and crop regulation, planting of regular bearing varieties, maintenance of the population of pollinators and 'pollinizers' varieties in the orchard could ensure fruitfulness in crops. In this way, problem of unfruitfulness could be minimized in future by applying the need based corrective measures in package of practices of a particular fruit crop.

References

Bose, T.K. 1985. Fruits of India: Tropical and Subtropical. Naya Prokash, India.

Early, M., Bamford, K. and Adams, C. 2013. Principles of Horticulture. Elsevier Science, United Kingdom.

Hancock, J.F. 2008. Temperate Fruit Crop Breeding: Germplasm to Genomics. Springer Netherlands.

Palaniappan, 2001. Germplasm screening for salinity stress in tropical fruit species. Regional Training Course "Characterization, Evaluation and Conservation of Tropical Fruits Genetic Resources", organized by IPGRI, ICAR and IIHR.

Sharma, K.K. 2014. Soil and Orchard Management. Daya Publishing House, India.

Sharma, R.R. 2007. Fruit Production: Problems & Solutions. International Book Distributing Company, India.

Verma, L.R. and Jindal, K.K. 2004. Fruit Crops Pollination. Kalyani Publishers, India.

14

Post Harvest Management of Horticultural Crops

Shailendra K Dwivedi, A. K. Pandey
Ankit Pandey and Vivek Tiwari

Fruits and vegetables account for nearly 90% of total horticulture production in the country. India is the second largest producer of fruits and vegetables in the world and the leader in several horticultural crops namely Mango, Banana, Papaya, Cashewnut, Arecanut, Potato and Okra. However, the nature of horticultural crops being such that it's not easy to make assessment of their production. These crops, especially vegetables are grown in small plots, fields or in the backyard of the houses; do not have single harvesting in most of the cases which makes their assessment difficult. Many horticultural crops have multiple pickings in a single season. Similarly many fruit trees are scattered, which do not count for assessment.

India has witnessed increase in horticulture production over the last few years. Significant progress has been made in area expansion resulting in higher production. Over the last decade, the area under horticulture grew by 2.6% per annum and annual production increased by 4.8%. During 2017-18, the production of horticulture crops was 311.71 Million Tonnes from an area of 25.43 Million Hectares. The production of vegetables has increased from 101.2 Million Tonnes to 184.40 Million Tonnes since 2004-05 to 2017-18 and production of fruits has increased from 50.9 Million Tonnes to 97.35 Million Tonnes since 2004-05 to 2017-18.

Fruits and vegetables provide health benefits and are important for the prevention of illnesses. Fruits and vegetables contain a variety of nutrients including vitamins, minerals and antioxidants. Eating the recommended amount of fruits and vegetables each day can reduce the risk of chronic diseases. The healthiest choices are fresh fruits or frozen without added sweeteners. Fruits are naturally low in fat, sodium and calories, and rich in potassium, fiber, vitamin C and folate. Some high potassium fruits include peaches, cantaloupe, honeydew, oranges and bananas. Fiber in fruits helps to protect against heart disease and lower the cholesterol level. Vitamin C in foods like citrus and strawberries helps with wound healing and keeps gums and teeth healthy. Vegetables are rich in vitamin A, vitamin C, folate, fiber and potassium. Folate helps the body form red blood cells. It is especially important for women of childbearing age to consume folate-rich foods such as bell peppers, tomatoes and spinach to prevent neural-tube defects in babies. Vitamin A-rich foods such as sweet potatoes, carrots and butternut squash help keep your skin and eyes healthy and protect against infections. Eating produce can cut the risk of cardiovascular disease, stroke and type 2 diabetes and protect against various forms of cancer. The high level of fiber in fruits and vegetables can reduce chances of developing coronary heart disease. Eating potassium-rich foods such as bananas and potatoes can help reduce blood pressure, decrease bone loss and prevent development of kidney stones. Along with several health benefits, eating fruits and vegetables can make weight management easier, these are low in calories compared to other foods, so filling up on these foods can aid in weight loss or maintenance. Fruits and vegetables work as excellent substitutes in different recipes.

Fruits and vegetables production have increased in recent years due to the economic and nutritional importance. This increase is made possible by the numerous research advances made along the entire value chain. However, scientific research has been focussed mainly on production whilst neglecting post-harvest issues. Fruits and vegetables producers have therefore enjoyed good harvests in recent times, though the good harvests of those from developing countries do not translate into profit as most are lost after harvest. It has been revealed that the post-harvest quality and shelf-life of fruits and vegetables depend on some post-harvest handling practices and treatments carried out after harvest. Handling practices like harvesting, precooling, cleaning and disinfecting, sorting and grading, packaging, storing, and transportation play an important role in maintaining quality thereby extending shelf-life.

Principles of harvesting

- Harvest should be completed during the coolest time of the day, which is usually in the early morning or late evening, and produce should be kept under shade.
- The produce has to be handled gently.

- Crops destined for storage should be as free as possible from skin breaks, bruises, spots, rots, decay, and other deteriorations.
- Bruises and other mechanical damage not only affect appearance, but provide entrance to decay organisms as well.

Harvesting of fruits and vegetables

Harvesting is the process of detaching a produce from the mother plant at the proper stage of maturity by an appropriate technique and as rapidly as possible with minimum damage or loss imparted to the commodity all at a relatively low cost. When quality is the primary consideration, not the cost, not the speed, and not the time, hand harvesting is the best approach. Generally, fruits intended for fresh market are hand harvested. Harvesting also bring out wound responses like ethylene production and increase respiration in the tissues. Mature tissues generally show only small responses to harvesting because they store carbohydrates reserves and relatively low respiration and transpiration rates, and its destined for natural separation by abscission any way. Rapidly metabolizing tissues such as leafy vegetables/immature fruits & vegetables exhibits larger responses to harvesting.

Time of harvesting also affects quality of fruits. Fruits harvested before 10 AM in the morning and transported to pack house for sorting, grading, and packing yield better quality and lasts longer. It is desirable that the fruits are harvested during the cooler parts of the day to reduce the risk of heat injury and sunburn. Therefore, morning harvesting and within 10 AM transportation to destination pack house or market is always preferred in order to control damage due to high temperature. In case of grapes harvesting in India, it starts at 6 o'clock in the morning and harvested produce reach pack house by 10 AM. It facilitates faster pre-cooling also and yield better quality.

METHODS OF HARVESTING

Selection of suitable method for harvesting of the produce is necessary otherwise bruises or injuries during harvesting may later manifest as black or brown patches making them unattractive. Latex coming out of stem in mango should not be allowed to fall on fruits as it creates a black spot. Injury to peel may become an entry point for microorganisms causing rotting. Some harvesting gadgets like mango and other fruit harvester have been developed by different ICAR stations. There are basically three methods most commonly used for harvesting any fruit.

i) Harvesting of individual fruits with hand by pulling or twisting the fruit pedicel

ii) Harvesting individual fruits or fruit bunch with the help of fruit clippers/ secateurs/scissors

iii) With harvester specially designed for harvesting. One important demerit of this method is pulling of little peel along with pedicel end renders the fruits for quick spoilage.

Hand Harvesting

Hand harvesting has several advantages, especially with reference to selective picking at the right stage of maturity to allow for the maximum quality development in the fruits or vegetables prior to harvest (especially if the maturity stage is visually and easily assessable), with minimum damage done to the produce during picking and transfer to the appropriate container for subsequent handling, storage and transportation. In addition, it involves minimum or almost no capital investment and the harvest output is proportional to the number of people at work. The only capital required is perhaps shelter and transpiration for the workers and small harvesting tools to separate the produce from the mother plant. One can supply more or fewer people based on needs. The main problem with hand harvesting centers around the workers needed to do the job. There could be an acute shortage of labor when the need is the greatest during the harvesting season. The seasonal nature of fruit and vegetable crops makes it difficult to keep the entire work force engaged throughout the year. So is hand harvesting. Quality of fruits and vegetables is very important for successful marketing. It can only be satisfactorily obtained by hand harvesting at the right stage. Hence, even today, most of the fruits and vegetables intended for fresh market are almost entirely hand harvested.

Benefits of hand harvesting

- Hand harvesting is less expensive.
- Less damage and harvest rate (times) can be increased.
- The main benefit of hand harvesting over mechanized harvesting is that humans are able to select the produce at its correct stage of ripening and handle it carefully. The result is a higher quality product with minimum damage. Examples,
- Breaking off – twisting off pineapple, papaya, tomato
- Cutting – snipping off mandarins and table grapes with secateurs and apple, roses etc
- Harvesting methods is also use full reducing incidence of fungal infection in papaya/grapefruit.-When fruit are cut from the tree using clipper shows less infection then the harvesting by twisting and pulling.

Demerits

Harvesting small fruits and from thorny plants are major obstacle (disadvantage).

Mechanized Harvesting

Mechanical harvesting is employed for the majority of fruits and vegetables intended for processing where they normally are converted to other forms, where physical appearance is not a major consideration, where the commodity is consumed within a short time after harvest so the quality may not deteriorate seriously in spite of some damage induced by the mechanical harvesting techniques, and where the commodity is needed in large quantities. Some advantages of mechanized harvesting are: Speed of harvest, Improved conditions for the workers, Reduced labor related problems etc.

The disadvantages and problems associated with mechanical harvesting systems are summarized below:

- Physical/mechanical damage to the crop
- Non-selectivity
- Separation of plant debris
- Damage to fruit trees during harvesting
- Large volume output
- Expensive machinery
- Social impact

Importance of Postharvest Technology

The Indian Economy is heavily dependent on agriculture, and this is projected to continue in the near future as well. The Agriculture and Allied Sector contributed approximately 13.9% of India's GDP and engages about 50% of the workforce. India is a rich horticultural country producing wide variety of fruits, vegetables, spices, ornamental and medicinal plants. India is second largest producer of fruits and vegetables in the world. Food production has been steadily increased in India due to advancement in production technology, but improper post-harvest management, processing, value addition and storage results in high losses in agricultural produces. Unfortunately, having such a huge production, a considerable postharvest loss to the tune of 10-25% occurs annually in vegetables mainly due to inefficient postharvest management practices. Processing of food products mainly as part of a cottage industry has been a long established traditional practice imbibed in many cultures of the country. However, with changing lifestyle patterns, increasing income, increasing preference toward, Ready to Eat and packaged foods, the significance of the Food Processing Industry has increased enormously. The term postharvest losses are defined as "losses that occur after harvest till the produce reaches consumers. It can quantity as well as qualitatively losses. Post-harvest losses are more painful and costlier than pre-harvest in terms of money and labour. Vegetables are highly perishable having moisture content of (80-90%). They are live commodities and continue their life processes like respiration and transpiration even after harvest. Water loss or transpiration is a major factor affecting quality of fruits and vegetables. In addition to lower saleable weight, loss of water can affect quality in many ways, including wilting, shrivelling, flaccidness, soft texture and loss of nutritional value. About 10-15% fresh fruits and vegetables shrivel and decay, lowering their market value and consumer acceptability. Minimizing these losses can increase their supply without bringing additional land under cultivation. Improper handling and storage cause physical damage due to tissue breakdown. Mechanical losses include bruising, cracking, cuts, microbial spoilage by fungi and bacteria, whereas physiological losses include changes in respiration, transpiration, pigments, organic acids and flavour.

Horticultural commodities are highly perishable in nature thus there may be a glut of fruits and vegetables in the market during the peak harvest season. These crops undergo a rapid transformation between the harvest and consumption which results spoilage and reduces market value. The spoilage has been estimated to be nearly 30-40 per cent in most of the produce which account for more than 25,000 crores of rupees every year. This is not only a loss to the growers but a net loss of huge human nutrition and wastage of inputs involved. These losses can be minimized to a considerable surplus with timely and safe management of post harvest produce.

The postharvest management of fruits and vegetables includes pre and post harvest practices, their harvesting, handling, packaging, storage, distribution, marketing, etc. Since fruits and vegetables contain a very high percentage of their fresh weight as water. Consequently, fruit exhibit relatively high metabolic activity when compared to other plant derived foods such as seeds. This metabolic activity continues post harvest and thus makes most fruits highly perishable commodities. This perishability, with its inherent short shelf life, that presents the greatest problem to the successful transportation and marketing of fresh fruits and vegetables. Thus, enhancement of their shelf life would be of great help in reducing postharvest losses, avoiding gluts in the peak season and avoiding distress sale. This would also help in ensuring more availability of fruits and vegetables without bringing additional land into production and fetching higher economic returns to the farmers.

The advantages of reduction in post harvest losses are as under

- Availability of fruits for longer durations.
- Better returns to the fruit grower
- Increase production without bring any additional land into fruits
- Availability of fruits to the consumers at low price
- Better nutrition to the consumers

Fruits are perishable in nature as it contains 80-95% moisture, high rate of respiration, more surface area and delicate texture. Lack of harvesting equipments, collection centers in major producing areas, suitable containers, commercial storage plants and lack of cold chain in complete post harvest handling are the possible reasons for providing a suitable media for the rotting and senescence.

Control Measures

- Growing of varieties having better shelf-life.
- Adoption of proper/ recommended cultural practices for the production quality fruits free from insect pest.
- Application of pre and post harvest treatment for the production of better quality fruits.
- Integrated approach to post harvest operations like harvesting at proper maturity, grading, sorting, removal of field heat, use of appropriate packing materials.
- Appropriate packaging
- Efficient and effective transportation according to the domestic or export purpose.
- Better storage and marketing.

Post Harvest Losses

Proper methods of processing, storage, packaging, transport and marketing are required for export of crops such as jute, tea, cashew nuts, tobacco, mango, litchi, nut, spices and condiments. One of the attributes to this post harvest system, as it is now constituted, is the large amount of wastage it involves. In case of food grains, some estimates suggest that in developing countries as much as $1/4^{th}$ to $1/3^{rd}$ of total crop may be lost as a result of inefficiencies in the post harvest system.

Losses of food crops refer to many different kinds of loss produced by a variety of factors. These include weight loss, loss of food values, loss of economic value, loss of quality or acceptability and actual loss of seeds themselves.

Horticultural produce is biological entity with various physiological activities like transpiration and respiration continuing even after harvesting. This process leads to the bio-chemical breakdown and cause spoilage of the produce. Spoilage is initiated by enzymes present inside the produce, involvement of micro organism, infestation of insect-pest and invasion of pathogens. By taking care of these factors, food products can be stored for longer period.

Food preservation is a form of processing of food to prevent it from spoilage and making it possible to store in a fit condition for future use.

Causes of Post harvest Losses

Although it is simple to suggest minimizing losses during various postharvest operations, achieving the goal is quite challenging. In order to do so, one must first understand the various causes of postharvest spoilage of fruits and vegetables and the factors that influence them, and secondly, use the postharvest conditions/ operations that will result in extending the shelf-life of the produce. The different causes of postharvest food losses may be broadly grouped as primary and secondary:

a) Primary Causes

- Biological and microbiological: Consumption or damage by insect-pests, animals and microorganisms (fungi and bacteria).
- Chemical and biochemical:
- Undesirable reactions between chemical compounds present in the food such as browning, rancidity, enzymatic changes, etc.
- Mechanical: Spillages, damages caused by abrasion, bruising, crushing, puncturing, etc.
- Physical: Improper environmental and storage conditions (temperature, relative humidity, air speed, etc.)
- Physiological: Sprouting, senescence, other respiratory and transpiratory changes.
- Psychological: Human aversion or refusal due to personal or religious reasons.

Many of these factors have synergistic effects, and the losses can greater with a combination of factors. For example, chemical, microbial, biochemical, or physiological activity in a stored product is significantly influenced by the storage conditions, especially temperature. A ten-degree change in temperature can result in a two- to three-fold change in these activities. This is a key factor utilized for advantage in cold and controlled atmosphere storage applications where the produce is held at the lowest possible temperature without getting into problems of chilling injury or freezer burn. On the other hand, if proper precautions are not taken in handling and transportation, increased product temperatures may result in a very rapid quality loss. Another example is the presence of mixed loading in a storage chamber. The emanations, especially trace gases like ethylene, from one product, may trigger deteriorative or ripening reactions in another. Again, the presence of ethylene is a factor that may have an advantage in controlled ripening chambers, but must be prevented in other situations.

b) Secondary Causes of Losses

Secondary causes usually are the result of inadequate or nonexistent input and may lead to conditions favorable for primary causes. This can include: improper harvesting and handling; inadequate storage facilities, inadequate transportation, inadequate refrigeration and/or inadequate marketing system. The various causes of postharvest spoilage also can be grouped based on the nature of biological and environmental factors (Kader, 1985).

The biological factors include

1. **Respiration:** Respiration is a process by which all living cells break down organic matter into simple end-products with release of energy and CO_2. The result is loss of organic matter, loss of food value and addition of heat load which must be taken into account in refrigeration considerations. The higher the respiration rate of produce, the shorter is its shelf-life.
2. **Ethylene production:** Ethylene has a profound effect on physiological activities. Used in ripening chambers, it can trigger physiological activity even in trace amounts. Most living commodities produce ethylene as a natural product of respiration.
3. **Compositional changes:** Many changes occur during storage, some desirable and some undesirable. For example, loss of green color is desirable in fruits but not in vegetables. Development of carotenoid pigments may have nutritional importance. There will be changes in carbohydrates, proteins, and all other food components.
4. **Growth and development:** In most produce there is continued growth and development even after harvest. Characteristic activities are sprouting

of potatoes, onions and garlic, elongation of asparagus, seed germination in fruits like tomatoes, lemons, etc.

5. **Transpiration:** Transportation refers to water loss resulting in shriveling and wilting due to dehydration and is undesirable due to loss of appearance, salable weight, texture and quality.

6. **Physiological breakdown:** This includes freezing injury or frost damage in commodities subjected to temperatures below their freezing point which can occur in the field or during transportation/storage. Chilling injury is mainly associated with tropical and sub-tropical commodities held for prolonged periods at temperatures between 5°C and 15°C. Heat injury can result in commodities exposed to direct sunlight or excessively high heat for prolonged intervals.

7. **Other factors:** These include physical/mechanical damage to the produce occurring during harvesting, handling, storage and transportation, as well as spoilage due to pathological causes (attack by microorganisms such as bacteria and fungi). The environmental factors include temperature, relative humidity, atmospheric composition, light and other factors (fungicides, growth regulators, etc). It generally is recognized that higher temperature will result in increased respiratory activity and hence lowered shelf-life. Very high relative humidity conditions may lead to mold growth on produce surfaces while lower relative humidity can result in desiccation. Lowering of oxygen and increasing of carbon dioxide levels in storage atmospheres have been successfully used to promote micro respiration in produce and thus extend the shelf-life. The various causes of postharvest losses will be discussed in greater detail in the forthcoming chapters, which also cover measures that can be taken to minimize these deteriorative changes. This chapter serves to provide an overview of the various causes of postharvest spoilages. Basically, one should try to minimize losses due to each and every factor in order to extend the duration of postharvest storage. The best technique would certainly involve harvesting the produce at the optimum stage of maturity, followed by quick cooling, packaging and transfer to controlled atmosphere storage, where the temperature, relative humidity, air velocity and atmospheric composition are set at the most appropriate level for the produce in question. The produce would be ideally left in this primary storage almost until ready for the final shipment to the retailer for quick transfer to the consumer to get at least a week's high-quality life for the produce under in-home refrigerated-storage conditions. They all sound relatively simple, but the fact that the harvesting, precooling, storage, transportation, packaging and handling requirement for each commodity can be different makes the postharvest handling system very complex. Classification of fruits and

vegetables according to principal causes of postharvest losses and poor quality and in order of importance are given into table 1.

Table 1. Classification of fruits and vegetables according to principal causes of postharvest losses and poor quality and in order of importance.

Group	Examples	Principal causes of postharvest losses and poor quality in order of importance
Root Vegetables	Carrot	Mechanical injuries
	Beet	Improper curing
	Onion	Sprouting and rooting
	Garlic	Water loss (shriveling)
	Potato	Decay
	Sweet Potato	Chilling injury (subtropical and tropical root crops)
Leafy vegetables	Lettuce	Water loss (wilting)
	Chard	Loss of green color (Yellowing)
	Spinach	Mechanical injuries
	Cabbage	Relatively high respiration rates
	Green onion	Decay
Flower vegetable	Artichoke	Mechanical injuries
	Broccoli	Yellowing and other discolorations
	Cauliflower	Abscission of florets
		Decay
Immature-fruit vegetables	Cucumber	Over maturity at harvest
	Squash	Water loss (shriveling)
	Eggplant	Bruising and other mechanical injuries
	Peppers	Chilling injury
	Okra	Decay
	Snap beans	
Mature-fruits vegetables and fruits	Tomato	Bruising
	Melons	Over-ripeness and excessive softening at harvest
	Citrus	Water loss
	Banana	Chilling injury (Chilling sensitive fruits)
	Mangoes	Compositional changes
	Apples	Decay
	Grapes	
	Stone fruits	

Source: Kitinoja and Kader, 2002

Post Harvest Operations

Post-harvest handling operations encompass those steps required for harvested produce to be selected or modified to meet market quality standards and to be packaged in a form suitable for storage or marketing. Although post-production operations vary from country to country and region to region throughout the world, procedures are similar among the developing countries. However, operations diversify with farm size such as small landholders, medium scale farmers and progressive growers. Post-production operations will be dissimilar between the developed and developing countries. Different post harvest operations of horticultural commodities are described below in Figure 1.

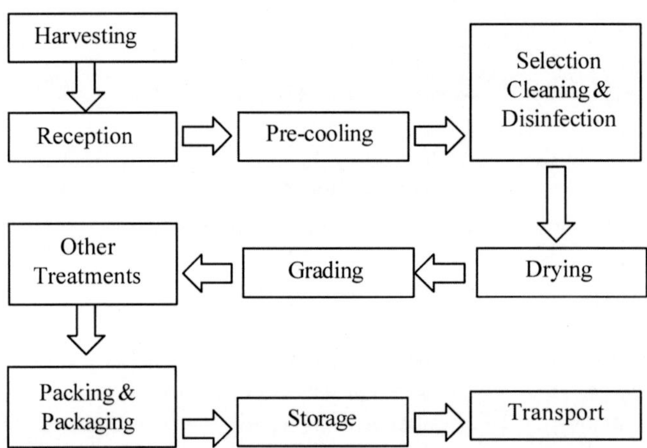

Figure 1: Different unit operation involved in fruits and vegetables

A. Precooling

Pre-cooling is the key component in the preservation of quality for perishable fresh produce in post-harvest systems. Pre-cooling is also very closely linked to the other operations such as handling and storage.

Pre-cooling is the rapid removal of heat from freshly harvested produce. This process is typically done before the produce is shipped to market or put into cold storage.

There are a variety of precooling techniques available for use in the horticultural industry. The principal methods of precooling highly perishable produce include room cooling, hydro cooling, forced air cooling, package icing, vacuum cooling and cryogenic cooling, with many variations and alterations within these techniques. In general most of the cooling is done at the packing houses or in central cooling facilities. The temperature is the most important factor affecting the post harvest life and quality of fruits and vegetables. Quality loss after harvest occurs as a result of physiological and biological processes, the rates of which are influenced primarily by product temperature. As the maintenance of market quality is of vital importance to the success of the fruits and vegetables, it is necessary not only to cool the product but to cool it as quickly as possible after harvest. The process of precooling is the removal of field heat which arrest the deteriorative and senescence processes so as to maintain a high level of quality that ensures customer satisfaction. The classification of crops according to their cooling methods is given in the table 2.

There are following types of precooling being used for horticultural produces.

i) **Room cooling:** Precooling produce in a cold-storage room or pre-cooling room is an old well-established practice. This widely used method involves the placing of produce inboxes (wooden, fibre board or plastic), bulk containers or various other packages into a cold room, where they are exposed to cold air. It has been shown that the liners in packages and paper curtains over the vegetable produce retard the cooling considerably. The best cooling rates to be achieved more space is required than for good storage management and thus some re-arranging of the produce after cooling may be necessary to utilize the storage space fully. In a study asparagus was room cooled to 8°C demonstrated that even through the use of this slow precooling technique an extension of saleable life by 2 days is achievable. Typically the cold air is discharged into the room near the ceiling, and sweeps past the produce containers to return to the heat exchangers. The cooled air is generally supplied by forced or induced draft coolers. Therefore, as to achieve fast and efficient cooling, care should be taken that the correct packaging (well vented) or containers and stacking patterns are used. Air velocities around the packages should be at least 60 m/min to provide the necessary turbulence to achieve heat removal and therefore attain sufficient cooling. As much of the cooling is achieved by conduction, room cooling gives a slow and variable temperature reduction; therefore perishable produce used in this method must be tolerant of slow heat removal. A conventional cold store is unsuited for this operation because as much as three-quarters of the refrigerator capacity may be required simply to remove field heat and the cooling rates are frequently no better than 0.5 °C/h. The rooms commonly used for highly perishable fruit are designed to have an airflow rate of about 170 to 225 m^3/min for a room with a capacity of 15,000 kg and sufficient refrigeration so as to cool the fruit to 5 °C in approximately 12 h. Containers are stacked individually so that cold air from the ceiling blows over or around the produce to contact all surfaces of the containers.

ii) **Forced air cooling:** Forced air cooling was developed to accommodate products requiring relatively rapid removal of field heat immediately after har vest. Forced air cooling pulls or pushes air through produce containers, greatly speeding the cooling rate of any type of produce. Many types of forced-air coolers can be designed to move cold air past the commodities. When utilizing a tunnel-type forced-air cooler, the canvas sheets must be well sealed over the top of the load, and pallet openings blocked for the cooler to function properly Package vents must be aligned between boxes to allow the air to flow across a pallet of boxes. Forced air cooling was developed by Guillou to accommodate products requiring relatively rapid removal of field heat immediately after harvest. Forced air or pressure cooling is a modification of room cooling and is accomplished by exposing packages of produce to higher air pressure on one side than on the other. It was indicated that for successful forced air cooling operations, it is required that containers with vent holes be placed in the direction of the moving air and packaging materials that would interfere with free movement of air through the containers should be minimized. The produce can be cooled by a variety of different forced air cooling arrangements.

These include

a) Air circulated at high velocity in refrigerated rooms,

b) By forcing air through the voids in bulk products as it moves through a cooling tunnel on continuous conveyors, and

c) By encouraging forced airflow through packed produce by the pressure differential technique.

The product cooling rate is affected by numerous variables and, therefore, the overall cost of the forced air-cooling will vary. These variables include product size and shape; thermal properties; product configuration (bulk or packaged); carton vent area; depth of product load during cooling; initial product temperature; final desired product temperature and air flow rate, temperature, and relative humidity. The cooling air comes in direct contact with the product being cooled and cooling is much faster than with conventional room cooling. This gives the benefit of rapid reduction in field heat and, therefore, quick product movement through the cooling plant. It was reported that it took 2 to 3 times longer than hydro cooling or vacuum cooling, while it was stated that cooling by the forced air method was usually 4 to 10 times faster than room cooling but that hydro cooling and vacuum cooling was 2 to 23 times faster than forced air cooling. Researchers were examined the energy use of a variety of precooling techniques and found that forced air coolers gave the lowest efficiency due to high levels of heat input as a result of improper design and operation of the air moving system. It has been explained that when very rapid cooling is required, forced air cooling is more costly than other precooling methods. Forced air cooling is also used for a variety of vegetables including broccoli, cauliflower, green beans, celery, cucumber, mushrooms and tomatoes.

Table 2. Classification of crops according to their cooling methods

Room cooling	Forces-air cooling	Hydro-cooling	Ice cooling	Vacuum cooling
Artichoke	Avocado	Artichoke	Belgian endive	Belgian endive
Banana	Banana	Asparagus	Broccoli	Brussels sprouts
Beans (dry)	Barbados cherry	Beet	Brussels sprouts	Carrot
Beet	Berries	Belgian endive	Cantaloupe	Cauliflower
Breadfruit	Brussels sprouts	Broccoli	Chinese cabbage	Chinese cabbage
Cabbage	Cactus leaves	Brussels sprouts	Carrot	Celery
Cactus leaves	Cassava	Cantaloupe	Escarole	Escarole
Cassava	Coconut	Cauliflower	Green onions	Leek
Coconut	Cucumber	Carrot	Kohlrabi	Lettuce
Custard apple	Eggplant	Cassava	Leek	Lima bean
Garlic	Fig	Celery	Parsley	Mushrooms
Ginger	Ginger	Chinese cabbage	Pea/snow peas	Snap beans
Grapefruit	Grape	Cucumber	Spinach	Snow peas
Horse radish	Grapefruit	Eggplant	Sweet corn	Spinach
J. artichoke	Guava	Escarole	Swiss chard	Sweet chard
Kohirabi	Kiwi fruit	Green onions	Watercress	Swiss chard
Kumquat	Kumquat	Horse radish		Watercress
Lime	Lima bean	J. artichoke		
Lemon	Mango	Kiwi fruit		
Melons	Melons	Kohlrabi		
Onion	Mushrooms	Leek		
Orange	Okra	Lima bean		
Cucumber	Orange	Orange		
Pineapple	Papaya	Parsley		
Potato	Passion fruit	Parsnip		
Pumpkin	Pepper (Bell)	peas		
Radish	Pineapple	Pomegranate		
Summer squash	Pomegranate	Potato (early)		
Sweet potato	Prickly pear	Radish		
Tomato	Pumpkin	Snap beans		
Turnip	Snap beans	Snow peas		
Watermelon	Snow peas	Spinach		
	Strawberry	Summer squash		
	Summer squash	Sweet corn		
	Tangerine	Swiss chard		
	Tomato	Watercress		

Source: Sargeant *et al.*, 2000

iii) **Hydrocooling:** Hydro cooling has been used since 1923 when it was developed as an out growth of celery washing. It is because of its simplicity and effectiveness that hydrocooling is a popular precooling method. Hydro cooling essentially is the utilization of chilled or coldwater for lowering the temperature of a product in bulk or smaller containers before further packing. Various types of hydro cooler are available, some of which include conventional (flood) type, immersion type, and batch type. The flood type hydro cooler cools the packaged product by flooding as it is conveyed through a cooling tunnel. With the batch system, chilled water is sprayed over the product for a certain length of time, depending on the season and the incoming product temperature. These hydro coolers have a smaller capacity than conventional hydro coolers and are therefore less expensive. The immersion type cooler uses a combination of immersion and flood cooling. Loose produce is immersed in cold water, and remains immersed until an inclined conveyor gradually lifts the products out of the water and moves it through an over head shower. It is nearly twice as rapid as conventional hydro cooling methods, due to the fact that moving chilled water completely surrounds the exterior surface of the produce and hence facilitates quicker temperature reduction. The bulk type cooler has the added benefit over the flood type cooler of allowing greater packaging flexibility. Many vegetables are successfully hydrocooled such as sweet corn, celery, asparagus, radishes and carrots. Main benefits of hydro cooling are that it is seen to prevent loss of moisture during the cooling process. Another advantage of this technique is that it is very rapid in contrast to other precooling techniques available. Field heat can be removed in 20-30 min using hydrocooling instead of several hours normally needed for forced air-cooling. In a study it was demonstrated that the capacity of hydrocooling may be at least twice as high as that of air cooling. Hydro cooling with ice and $CaCl_2$ for 30 min produces the best storage characteristics of tomatoes resulting increase in shelf-life up to 13 days as compared to air cooling for 24 hours, hydro cooling with raw water for 45 min, hydro cooling with ice water for 30 min. At typical flow rates and temperature differences, water removes heat about 15 times faster than air. It is because of their relatively high cost, hydro coolers must be operated for considerable periods each year to be economically justified.

iv) **Contact or package icing:** Before the advent of comparatively modern precooling techniques, contact or package icing was used extensively for precooling produce and maintaining temperature during transit. Crushed or flaked ice for package icing can be applied directly or as slurry in water. Package ice can be used only with water tolerant, non-chilling sensitive products and with water tolerant packages. In liquid icing, ice slurry is used instead of plain crushed ice as it can sustain cooling requirements better.

The major advantage of icing is that produce does not dry as it is cooled. It has been reported that water loss from broccoli precooled by hydro cooling and top icing gave similar results. Another advantage is that in addition to removing field heat, package icing can maintain low product temperature during transit and therefore refrigerated transportation may not be necessary for short transport duration. Although icing requires relatively small outlays of special equipment, a large weight of ice must be shipped, thus increasing costs, and also water-proof containers which are more expensive than normal are required for this cooling technique. Icing can be effectively used to cool products such as collards, kale, Brussels sprouts, broccoli, radishes, carrots and onions.

v) **Vacuum cooling:** Rapid cooling of horticultural produce can be carried out with vacuum cooling. Vacuum cooling is achieved by the evaporation of moisture from the produce. The evaporation is encouraged and made more efficient by reducing the pressure to the point where boiling of water takes place at a low temperature. The heat required to vaporize this water is removed from the product surface, hence the cooling rate is limited by heat and mass transfer, i.e. evaporation rate of water from the products surface and inner tissues. Therefore the rate of cooling depends primarily upon the ratio of surface area of the product to its weight or volume, the ease with which water is given up from the product tissues, the rate of vacuum drawn in the flash chamber and the temperature of the load at the start.

vi) **Cryogenic cooling:** The use of the latent heat of evaporation of liquid nitrogen or solid CO_2 (dry ice) can produce 'boiling 'temperatures of -196°C and -78°C, respectively. This is the basis of cryogenic precooling. In cryogenic cooling, the produce is cooled by conveying it through a tunnel in which the liquid nitrogen or solid CO_2 evaporates. However, at the above temperatures the produce will freeze and thus be ruined as a fresh market product. This problem is prevented by careful control of the evaporation rate and conveyor speed. Cryogenic cooling is relatively cheap to install but expensive to run. Its main application is in cooling crops such as soft fruits, which have a seasonal production period.

It is essential part of the proper temperature management of fruits and vegetables. Precooling is essential for the removal of heat or the reduction in temperature of the perishable produce as soon as possible after harvest. This process slows the respiration rate and minimizes other deteriorative processes and thus helps to maintain quality at a high level. Precooling in conjunction with the proper storage or transportation allows for the extension of shelf-life of the horticultural produce which results in more satisfied customers at all levels of purchase. Within precooling a variety of different techniques exist for fruits and vegetables. Hydro cooling, vacuum cooling, room cooling, icing, forced air cooling and cryogenic cooling are the principal methods in commercial use at present. Each of these individual techniques also has many variations, leading to a great diversity of perishable produce which may be pre cooled. As consumer awareness and sophistication are ever increasing due to the growing fear of chemical residues and the uncertainty surrounding genetically modified foods presently, and with the change to organic products continuing, alternative techniques of extending shelf-life and maintaining high level of quality are being investigated. Precooling is one of the techniques which adheres to this ethos and should be applied widely for most fruits and vegetables to attain its true potential.

B. Sorting and Grading

Cleaning and grading are important post harvest operations undertaken to remove foreign and undesirable matters. From the threshed crops/grains/seeds and to further separate the grains/products into various fractions.

a) Sorting

Sorting, like grading, facilitates subsequent processing operations, such as peeling, pitting, blanching, slicing, and filling of containers. It is beneficial for heat and mass transfer operations, where processing time is a function of the size of the product (e.g. heat conduction, mass diffusion). Separating the different types of fruits from fruit lot is called sorting.

Fruits and vegetables are sorted on the basis of colour, damage and size. Mostly sorting on the basis of colour and damage is done manually, but electronic eye has been used successfully in the pilot studies, and its future general application appears promising. The shape of the fruits and vegetables should be suitable for mechanical handling and processing. The size and shape of some fruits pose problems in processing operations, e.g. apples, mangoes, and papayas.

Some of the equipment used for sorting of fruits and vegetables are as follows

1. Screens
2. Diverging belts
3. Roller sorters
4. Weight sorters

   It depends on

1. Damage product: Damaged product should be avoided. It should be separated from fruit lot and well mature product should taken.

2. Diseases: Diseased free fruit and vegetable should be collected from a lot. It is very important for sorting.
3. Insect cutting products: Insect cutting fruit must be avoided because consumer wants good fruit.
4. Maturing: Mature fruit is very tasty. So maturity is very much needed for fruit.
5. Color: Colours should be lightable. Well coloured fruit should be selected that attract consumer. Colour also indicates the maturity level of seed.
6. Shape: In case of packaging, shape is essential. Some type of fruit should be selected for packaging in case of sorting.
7. Size: Size also an important matter for sorting fruits from a fruit lot. We should selected same size of fruit.

b) Grading

Grading is sorting of fruits and vegetables into different grades according to their size, shape, colour, and volume to fetch high price in market.

Or, Arrangement of fruits into different groups by separating from a fruit lot is called grading. Grading depends on Cultivar, Size, Appearance, Colour and Quality.

Grading is the classification of materials on the basis of commercial value, end usage (product quality), and official standards. For, example, grading is necessary to avoid the further processing of blemished, spoiled, or products not meeting the quality requirements. Grading is done mostly by hand (e.g. inspection of fruits after washing), but when physical characteristics are also indicative of product quality, grading can be done through machinery. In rice, the white kernels are separated optically from the off colour grains or from foreign matter, and the lighter unripe tomatoes can be separated from the ripe according to their specific weight. Grading of fruits and vegetables after harvesting is an essential step in post-harvest management. Grading of fruits and vegetables on the basis of physical characteristics like weight, size, colour, shape, specific gravity, and freedom from diseases depending upon agro-climatic conditions. The known methods of grading of fruits and vegetables are manual grading, size grading.

Grading of fruits and vegetables in the fresh form for quality is essential, as the people are becoming quality conscious day by day. Further, upon arrival of fruits and vegetables at the processing centres, they should be graded strictly for quality. The immature, properly mature and over mature fruits and vegetable should be sorted out for the best attributes.

Advantages of Grading

- Losses the selling price due to presence of substandard products or specimen can be easily avoided.
- It increased marketing efficiency by facilitating buying and selling a produce without personal selection.
- Grading enhanced to set good price for graded products.
- Heavy marketing cost in packing and transportation can be avoided by grading.
- In grading, diseased and defected specimen are not damaged due to contact of diseased specimens and thus gets high price in market.
- By grading, there is fairness to both Buyers and Sellers.
- Properly graded vegetables and fruits are purchased by the consumer easily without inspection.

C. Cleaning and Washing

It is done to remove the dust and microorganisms from the surface of the fruit. The washing of fruits should be done under running water (chlorinated) only in those cases which has acquired latex stains e.g. mango and banana. It should be carried only if it is required absolutely and if it has to be done than fungicides should be applied immediately after washing. Washing cut produce with potable water may reduce microbiological contamination. In addition, it removes some of the cellular fluids that were released during the cutting process thereby reducing the level of available nutrients for microbiological growth. The following should be considered:

- Water should be replaced at sufficient frequency to prevent the build-up of organic material and prevent cross-contamination.
- Antimicrobial agents should be used, where necessary, to minimize cross-contamination during washing and where their use is in line with good hygienic practices. The antimicrobial agents levels should be monitored and controlled to ensure that they are maintained at effective concentrations. Application of antimicrobial agents, followed by a wash as necessary, should be done to ensure that chemical residues do not exceed levels as recommended by the Codex Alimentarius Commission.
- Drying or draining to remove water after washing is important to minimize microbiological growth.

Fruits and vegetables often contain a great diversity of microûora and are frequently involved in food-borne outbreaks. Since fruits and vegetables are mainly consumed uncooked or minimally-processed (such as in ready-to-eat salads), microbiological safety becomes an important issue to minimize consumers risks. Recently, a number of outbreaks have been traced to fresh-cut fruits and vegetables, which were caused due to inadequate sanitary conditions. The investigations of these outbreaks showed that the quality of water used for washing was crucial. It is well known that disinfection is one of the most critical processing steps in fresh-cut fruits and vegetable production. This step commonly affects the quality, safety and shelf-life of the end product. Washing is designed to remove dirt, selected pesticides and to detach microorganisms to enhance quality. Sanitization is the killing of contaminated microorganisms after washing. Chemical methods of cleaning and sanitizing involve the application of mechanical washing in the presence of sanitizers. Sanitizers can reduce the growth of natural microbial populations on the surface of fresh-cut produce by 2–3 log units and can reduce contaminated pathogens.

It is well known that water serves as a source of cross contamination as it is re-used. The water used for washing may result in the building of microbial loads if it is not properly managed. The importance of water quality used for washing need to be applied and sanitizing agents could be used to maintain the quality of the water. This can prevent cross contamination and microbial spoilage of the final product. In general, it could be assumed that the cleaning action of the washing removes microorganisms by detaching them from the products and sanitizing agent eliminates them by killing. Ideal sanitizing agent should have two important properties: a sufficient level of antimicrobial activity and a negligible effect on the sensory quality. A range of disinfectant and sanitizers are described in the next sections. These are chlorine, chlorine dioxide (ClO_2), organic acids, ozone, hydrogen peroxide, electrolyzed water and tri sodium phosphate (TSP).

D. Curing

Some vegetables like potato, sweet potato, yam, cassava, onion and garlic are stored better and for long duration when they are cured before storage. Curing is most simple and effective way of reducing water loss and decay during post-harvest storage. In curing bruised and injured surface are allowed to heal. During the process of curing, wounds are healed by producing a new cork cambium, thus preventing infection by pathogenic organism. Onions and garlic are cured to dry the necks and outer scales. For the curing of onion and garlic, the bulbs are left in the field after harvesting under shade for a few days until the green tops, outer skins and roots are fully dried. The optimum condition for curing for root, tuber, bulb vegetables are given in the table 3.

Table 3. Curing temperature and RH required for different crops

Commodity	Temperature	RH (%)	Days
Cassava	30-40	90-95	2-5
Onion and Garlic	35-40	60-75	4-5
Potatoes	15-20	85-90	5-10
Sweet Potatoes	30-32	85-90	4-7
Yam	32-40	90-100	1-4

E. Peeling

Peeling is one of the integral parts of a food processing, and the majority of agricultural crops need to be peeled in order to remove at the initial stage of food processing. Peeling removes inedible portion (peel, seeds, and stalk) of fruits and vegetables. However, the susceptibility to spoilage increases due to acceleration of physiological process and the exposure of the tissues to microorganisms. The shelf-life and quality of fresh-cut produces could be compromised with the peeling. The goals of optimum peeling operation are:

1. Minimizing product losses,
2. Types of products e.g., potato products,
3. Minimizing heat ring formation e.g., apple, potato,
4. Minimizing energy and chemical usage, and
5. Minimizing the environmental pollution

Peeling operation can be grouped under following categories

- Manual peeling (knife or blade),
- Mechanical peeling (abrasive devices, devices with drums, rollers, knifes or blades and milling cutters),
- Chemical peeling,
- Enzymatic peeling, and
- Thermal peeling (Flame or dry heat peeling, steam or wet heat peeling, thermal blast peeling, and vapour explosion or vacuum peeling.

F. Cutting

Cutting, slicing, dicing, and shredding are non-thermal food operation for size reduction. This process reduces the preparation time by consumers. Cutting removes inedible and discoloured portions from foods using knife, chopper and slicer. However, in this food processing injured tissues are removed so that these are unavailable for microbial spread. The cut tissues results in reduced respiration and enzymes activity, thus retarding rapid spoilage and increases shelf life. The cutting process accumulates fluids on the cut surface, increases microbial load and enzymes activity. Some of the potential risk of microbial contamination in horticultural commodities is tabulated in table 4 along with recommended preventive measures.

Table 4. Potential risk of microbial contamination in horticultural commodities and recommended preventive measures

Production Step	Risks	Prevention
Production field	Animal fecal contamination	Avoid animal access, wild, production or even pets.
Fertilizing	Pathogens in organic fertilizers	Use inorganic fertilizers. Proper composting
Irrigation	pathogens in water	Underground drip irrigation Check microorganisms in water
Harvest	Fecal contamination	Personal hygiene, Portable bathrooms. Risk awareness
	Pathogens in containers and tools	Use plastic bins. Cleaning and disinfecting tools and containers
Pack house	Fecal contamination	Personal hygiene. Sanitary facilities. Avoid animal entrance. Eliminate places may harbor rodents.
	Contaminated water	Alternative methods for pre-cooling. Use potable water. Filtration and chlorination of re-circulated water. Multiple washing.
Storage and transportation	Development of microorganisms on produce	Adequate temperature and relative humidity. Watch conditions inside packaging. Cleaning and disinfection of facilities. Avoid repackaging. Personal hygiene. Do not store or transport with other fresh products. Use new packing material.
Sale	Product contamination	Personal hygiene. Avoid animal access. Sell whole units. Cleaning and disinfection of facilities. Discard garbage daily.

Source: FAO, 2004

G. Waxing

Edible coatings are defined as the thin layer of material which can be consumed and provide a barrier to oxygen, microbes of external source, moisture and solute movement for food. In edible coating a semi permeable barrier is provided and is aimed to extend shelf life by decreasing moisture and solute migration, gas exchange, oxidative reaction rates and respiration as well as to reduce physiological disorders on fresh cut fruits. different types of Edible coating materials used for different horticultural commodities are being mentioned into table 5.

Quality retention is a major consideration in modem fresh fruit marketing system. Waxes are esters of higher fatty acid with monohydric alcohols and hydrocarbons and some free fatty acids. But coating applied to the surface of fruit is commonly called waxes whether or not any component is actually a wax. Waxing generally reduces the respiration and transpiration rates, but other chemicals such as fungicides, growth regulators, preservative can also be incorporated specially for reducing microbial spoilage, sprout inhibition etc. However, it should be remembered that waxing does not improve the quality of any inferior horticulture product but it can be a beneficial adjunct to good handling.

The advantages of wax application are

- Improved appearances of fruit.
- Reduced moisture losses and retards wilting and shrivelling during storage of fruits.
- Less spoilage specially due to chilling injury and browning.
- Creates diffusion barrier as a result of which it reduces the availability of O_2 to the tissues thereby reducing respiration rate.
- Protects fruits from micro-biological infection.
- Considered a cost effective substitute in the reduction of spoilage when refrigerated storage is unaffordable.
- Wax coating are used as carriers for sprout inhibitors, growth regulators and preservatives.
- The principal disadvantage of wax coating is the development of off- flavour if not applied properly. Adverse flavour changes have been attributed to inhibition of O_2 and CO_2 exchange thus, resulting in anaerobic respiration and elevated ethanol and acetaldehyde contents. Paraffm wax, Carnauba wax, Bee wax, Shellac, Wood resins and Polyethylene waxes used commercially.

Table 5. Edible coating materials used for different horticultural commodities.

S.No.	Fruit and vegetables	Used Edible coating
1.	Fresh-cut Apple	Sodium Alginate, Gellan Gum, Sunflower oil. Whey protein concentrate, WPI & WPP.
2.	Blueberry	CMC, Chitosan, Monoglycerdes.Sodium alginate, Calcium casienate.
3.	Tomato	*Aloe vera* gel
4.	Grapes	*Aloe vera* gel.
5.	Mango	Chitosan, *Aloe vera*.Trapica flour, Sago flour, Soy protein. Chitosan.
6.	Apple	Neem oil, Marigold flower extract, Guar gum & *Aloe vera*.Soybean gum, Paraffin wax, Jojoba oil & Arabic gum. WPC. SPI, Alginate, Carrageenan.
7.	Fresh-cut Pineapple	Pectin & Alginate.
8.	Strawberry	Sodium Alginate & Calcium Alginate gel.Arabic gum & Arjan gum Psyllium mucilage.Chitosan & Tragacenth.Linssed mucilage extract, Chitosan.Pectin, PVA, Starch, Soy protein, Gluten.
9.	Cantaloupe	*Aloe vera* gel.Pectin, Chitosan.
10.	Banana	PVA, CMC, Tannin.
11.	Guava (Fresh-cut)	*Aloe vera* juice.
12.	Pistachio	Chitosan and *Aloe vera* gel.
13.	Orange, Tomato, Mango, Papaya, Guava, Mushroom.	Veg. oil. Cellulose gum, Emulsifier.
14.	Potato	Chitosan, WPC, Coconut oil.
15.	Papaya	Chitosan, *Aloe ver.* Papaya leaf extract.
16.	Plum	CMC, Pectin.
17.	Sapota	*Aloe vera* Juice.
18.	Pumpkin, Carrot, Radish, Cantaloupe, Cucumber	Chitosan, calcium salt.
19.	Cucumber	Gum Arabic Powder. Beeswax, High Methoxy Pectin.

Source: Raghav *et al.*, 2016

H. Irradiation

Food Irradiation is a process whereby food is exposed to a carefully measured amount of intense radiant energy, called ionizing radiation. Irradiation is not only used to kill various germs in meat and meat products but also to reduce post harvest losses and increase shelf life of horticultural products. The scientists at Babha Atomic Research Centre (BARC) have shown that small doses of irradiation can inhibit sprouting in onion, potato and garlic. The ripening of mango, banana and other fruits and vegetables can also be delayed. Insect and their eggs which infest grains, pulses, vegetables and fruits can be killed. The first of these facilities is for spices irradiation and is functional at VASHI on the outskirts of Bombay. The department of Atomic energy has also established a food irradiator, which is named as POTON (for potato and onion) at Nasik in Maharastra. Sprouting of onion can be checked by gamma irradiation at a dose of 0.06 to 0.1 kgy. In potato, gamma irradiation at 0.1kgy can inhibit sprouting completely and also inhibit the light induced synthesis of chlorophyll and toxic alkaloid solanin. Laboratory studies have indicated that irradiated potatoes could be stored successfully for 6 months at 15°C with 10% loss. Fruits of climactric class such as banana, guava, mango and papaya where irradiation in the mature but unripe preclimactric stage is given in a close range of 0.25 to 0.75 kgy showed improved shelf life due to delay in the rate of ripening and senescence. Different doses of irradiation are being summarised in table 6.

Table 6. Dose requirement in various applications of food irradiation

Purpose	Dose (kGy)	Products
Low dose (up to 1kGy)		
a) Inhibition of sprouting	0.05-0.15	Potatoes, onions, garlic, ginger root etc.
b) Insect disinfestations and parasite disinfection	0.15-0.50	Cereals and pulses, fresh and dried fruits, dried fish and meat fresh pork, et.
c) Delay of physiological process (e.g. ripening)	0.50-1.0	Fresh fruits and vegetables
Medium dose (1-10 kGy)		
a) Extension of shelf-life	1.0-3.0	Fresh fish, Strawberries, etc.
b) Elimination of spoilage and pathogenic micro-organisms	1.0-7.0	Fresh and frozen seafood, raw or frozen poultry and meat, etc.
c) Improving technological properties of food	2.0-7.0	Grapes (increasing juice yield), dehydrated vegetables (reduced cooking time), etc.
High dose (10-50 kGy)		
a) Industrial sterilization mild (in combination with mild heat)	30-50	Meat, poultry, seafood prepared foods, sterilized hospital diets
b) Decontamination of certain food additives and ingredients	10-50	Spices, enzyme preparations, natural gum, etc.

Source: FAO, 1988

I. Packaging

Packaging plays a vital role in preserving food throughout the distribution chain. Without packaging, the processing of food can become compromised as it is contaminated by direct contact with physical, chemical, and biological contaminants. In recent years, the development of novel food packaging (modified atmosphere & active packaging) has not only increased the shelf life of foods, but also their safety and quality - therefore bringing convenience to consumers. Directly related, and interlinked, with food packaging is the concept of shelf life - the length of time that foods, beverages, pharmaceutical drugs, chemicals, and many other perishable items are given before they are considered unsuitable for sale, use, or consumption.

It is essential to minimize physical damage to fresh produce in order to obtain optimal shelf-life. The use of suitable packaging is vital in this respect. The most common form of packaging in this sector is the use of the ûbre board carton; however, for most produce, additional internal packaging, for example tissue paper wraps, trays, cups or pads, is required to reduce damage from abrasion. For very delicate fruits, smaller packs with relatively few layers of fruits are used to reduce compression damage. Moulded trays may be used which physically separate the individual piece of produce. Individual fruits may also be wrapped separately in tissue or waxed paper. This improves the physical protection and also reduces the spread of disease organisms within a pack. The storage of different horticultural commodities under refrigerator condition (5 °C) has direction correlation with repriration rate, which decides about the degratioon process of these commodities under such conditions (Table 7).

Table 7. Respiration rate of selected horticultural commodities

Commodities	Respiration Rate at 5 °C (mL CO_2lg^{-1}h^{-1})
Dried fruits and vegetables, nuts	<5
Apple, grape, potato (mature), onion, garlic	5-10
Apricot, carrot, cabbage, cherry, lettuce, peach, plum, pepper, tomato.	10-20
Blackberry, cauliflower, lima bean, raspberry, strawberry	20-40
Brussels sprouts, green onion, snap bean	40-60
Asparagus, broccoli, mushroom, pea, spinach, sweet corn.	>60

Source: Rizvi, 1981

Packaging can be a major item of expense in produce marketing, so the selection of suitable containers for commercial-scale marketing requires careful consideration. Besides providing a uniform-size package to protect the produce, there are other requirements for a container:

- It should be easily transported when empty and occupy less space than when full, e.g. plastic boxes which nest in each other when empty, collapsible cardboard boxes, fibre or paper or plastic sacks;
- It must be easy to assemble, fill and close either by hand or by use of a simple machine;

- It must provide adequate ventilation for contents during transport and storage;
- Its capacity should be suited to market demands;
- Its dimensions and design must be suited to the available transport in order to load neatly and ûrmly;
- It must be cost-effective in relation to the market value of the commodity for which used;
- It must be readily available, preferably from more than one supplier.

Owing to perishability, the bulk of fresh fruit and vegetables are still sold unpacked at retail or wholesale level in perforated polyethylene (PE) bags or nets. Long term transport of these materials must remain minimal to avoid brushing and consequent decay of the loosely package product. When packaging is required at the source or when an extended storage life is desired, the packaging ûlm should have a high gas permeability and anti-fog properties. The packaging of fresh vegetables and fruit provides the largest single use of printed PE bags. The pre-packaging of potatoes, carrots, onions, parsnips, beets, radishes etc. for market sales has become an important aspect of food distribution. Because of low water vapour permeability of PE and polypropylene (PP), both û lms and bags are sometimes perforated to allow the product to 'breathe'. Permeation is the key to extending fresh produce shelf life. For most commodities light is not an important inûuence in their post-harvest handling. However green vegetables, in the presence of sufficient light, could consume substantial amounts of CO_2 and produce O_2 through photosynthesis. Shock and vibration leads to damage to produce cells which causes an increase in respiration and may lead to enzymes being released that will cause browning reactions to begin.

References

Allende, A., Selma, M.V., Lo´pez-Ga´lvez, F., Villaescusa, R., and Gil, M. 2008. Role of commercial sanitizers and washing systems on epiphytic microorganisms and sensory quality of fresh-cut escarole and lettuce. Postharvest Biology and Technololgy, 49: 155-163.

Arfin, B.B. and Chau, K.V. 1988. Cooling of strawberries incar tons with new vent hole designs. AS HR AE Transactions, 92: 1415.

Chao-Chin, C., Tzou-Chi, H., Chen-Hsing, Y., Feng-Yi, S. and Ho-Hsien, C. 2011. Bactericidal effects of fresh-cut vegetables and fruits after subsequent washing with chlorine dioxide. International Conference on Food Engineering and Biotechnology. IPCBEE 9, IACSIT Press, Singapoore.

Corbo, M., Speranza, B., Campaniello, D., D'Amato, D. and Sinigaglia, M. 2010. Fresh-cut fruits preservation: current status and emerging technologies. In: Current Research, Technology And Education Topics In Applied Microbiology And Microbial Biotechnology (Mendez Vilas A., Ed.). Formatex Research Center, Badajoz.

FAO, 2004. Manual for the preparation and sale of fruits and vegetables: From field to market. FAO Agricultural Services Bulletin 151, Food and Agriculture Organization of the United Nations, Rome.

Gillies, S.L. and Toivonen, P.M.A. 1995. Cooling method influences the postharvest quality of broccoli. HortScience, 30: 313-315.

Horticultural Statistics at a Glance, 2018. Horticulture Statistics Division, Department of Agriculture, Cooperation & Farmers Welfare Ministry of Agriculture & Farmers Welfare Government of India.

Kader, A.A. 1985. Postharvest Biology and Technology: An overview. In: Postharvest Technology of Horticultural Crops (Kader A. A., Ed.), UC Publication No. 3311. University of California, Division of Agriculture and Natural Resources, Oakland.

Kitinoja, L. and Kader, A.A. 2002. Small-scale postharvest handling practices: A manual for horticultural crops, 4th Edition, Postharvest Horticulture Series No. 8E. University of California, Davis Postharvest Technology Research and Information Center, Davis.

Raghav, P.K., Agarwal, N. and Saini, M. 2016. Edible coating of fruits and vegetables: A review. International Journal of Scientific Research and Modern Education, 1(1): 2455-5630.

Rizvi, S.S.H. 1981. Requirements for foods packaged in polymeric films. Critical Reviews in Food Science and Nutrition, 14: 111-134.

Ryall, A.L. and Pentzer, W.T. 1967. The relation of airmovement, container type, and load arrangement to thecooling rate of fruits and vegetables. In: Proc. TwelfthInt. Cong. Refrig., 3: 87-92.

Sargent, S.A., Ritenour, M.A., Brecht, J.K. 2000. Handling, cooling, and sanitation techniques for maintaining postharvest quality. University of Florida, Coop Exten Serv, HS719. Available online at: http://www.gladescropcare.com/postharvest quality.pdf .

Scetar, M., Kurek, M. and Galic, K. 2010. Trends in fruit and vegetable packaging- A Review. Croatian Journal of Food Technology, Biotechnology and Nutrition, 5: 69-86.

Thompson, J.F. and Rumsey, T.R. 1984. Determining producttemperature in a vacuum cooler. ASAE paper No. 84-6-543.

Turk, R. and Celik, E. 1993. The effects of vacuum precoolingon the half cooling period and quality characteristic oficeberg lettuce. Acta Horticulturae, 343: 321.

WHO, 1988. Food irradiation: A technique for preserving and improving the safety of food. World Health Organization.

Zagory, D. 1999. Effects of post-processing handling and packaging on microbial populations. Postharvest Biology and Technology, 15: 313-321

15

Maturity Indices of Fruits and Vegetables

Pranava Pandey

The horticultural produce includes fruits, vegetables, flowers, plantation crops and spices. However, fruits and vegetables are the most consumed and important commodity among them. Morphologically and physiologically, the fruits and vegetables are highly variable and may come from a root, stem, leaf, immature or fully mature and ripe fruits. They have variable shelf life and require different suitable conditions during marketing. All fresh horticultural crops are high in water content and are subjected to desiccation (wilting, shrivelling) and to mechanical injury. Various authorities have estimated that 20-30% of fresh horticultural produce is lost after harvest and these losses can assume considerable economic and social importance.

Maturity

It is the state of reaching a stage of full or advanced development of tissue of fruits and vegetables, after which it will ripen normally. During the process of maturation, the fruit receives a regular supply of food material from the plant and at maturity, the abscission or corky layer formed at the stem end stops this inflow. Afterwards, the fruit depend on its own reserves, and carbohydrates are dehydrated and sugars start accumulating until the sugar acid ratio formation. In addition to this, typical flavour and characteristic colour also develop. It has been determined that the stage of maturity at the time of picking influence the storage life and quality of fruit, when picked immature like mango develop white patches or air pockets during ripening, thus lacks in in normal brix acid ratio or sugar acid ratio, taste and flavour. On the other hand, if the fruits are harvested over mature or full ripe, then they are easily susceptible to microbial and physiological spoilage and their storage life is considerably reduced. Such fruits continue to show numerous problems during handling, storage and transportation. Therefore, it is necessary or essential to pick up the fruits or vegetables at correct stage of maturity to facilitate proper ripening, distant transportation and to obtain maximum storage life.

Definitions related to maturity and ripening

i) **Mature**: It is that stage of fruit development, which ensures attainment of maximum edible quality at the completion of ripening process.

ii) **Maturation**: It is the developmental process by which the fruit attains maturity. It is the transient phase of development from near completion of physical growth to attainment of physiological maturity. There are different stages of maturation *e.g.* immature, mature, optimally mature, over mature.

iii) **Ripe**: This is the condition of maximum edible quality attained by the fruit following harvest. Only fruit which becomes mature before harvest can become ripe.

iv) **Ripening**: Ripening involves a series of complex physiological process like softening, colouring, sweetening and increase in aroma compounds during early stages of senescence of fruits in which structure and composition of unripe fruit is so altered that it becomes acceptable for consumption.

v) **Senescence**: Senescence can be defined as the final phase in the ontogeny of the plant organ during which a series of essentially irreversible events occur which ultimately leads to cellular breakdown and death.

There are different types of maturity as follows

i) Horticultural maturity

It is a developmental stage of the fruit on the tree, which will result in a satisfactory product after harvest.

ii) Physiological maturity

It refers to the stage in the development of the fruits and vegetables when maximum growth and maturation has occurred. It is usually associated with full ripening in the fruits. The physiological mature stage is followed by senescence.

iii) Commercial maturity

It is the state of plant organ required at a market. It commonly bears little relation to physiological maturity and may occur at any stage during development stage.

iv) Harvest Maturity

It may be defined in terms of physiological maturity and horticultural maturity, it is a stage, which will allow fruits / vegetables at its peak condition when it reaches to the consumers and develop acceptable flavour or appearance and having adequate shelf life.

Vegetables and fruits are harvested at harvest maturity stage, which will allow it to be at its peak condition when it reaches the consumers. It should be at a maturity that allows the produce to develop an acceptable flavour or appearance, it should be at a size required by the market, and should have an adequate shelf life. Time taken from pollination to horticultural maturity under warm condition, skin colour, shape, size and flavour and abscission and firmness are used to assess the maturity of the produce. The following examples are given for better understanding;

Skin colour - Loss of green colour in citrus and red colour in tomato.

Shape, size and flavour - Sweet corn is harvested at immature stage, smaller cobs marketed as baby corn. Okra and cow pea are harvested at mature stage (pre fiber stage). In chilli, bottle gourd, bitter gourd, cluster beans maturity is related to their size. Cabbage heads and cauliflower curds are harvested before the developemmnt of unpleasant flavour.

Abscission and firmness - Musk melon should be harvested at the formation of abscission layer. Cabbage and lettuce should be harvested at firmness stage.

Factors affecting maturity

1. **Temperature:** Higher temperature gives early maturity.
    - Gulabi (Pink) grapes mature in 100 days in Western India but only 82 days are enough in the warmer Northern India.
    - Lemon and guava takes less time to mature in summer than in winter. Sun scorched portions of fruits are characterized by chlorophyll loss, yellowing, disappearance of starch and other alcohol insoluble material, increase in TSS content, decrease in acidity and softening.
2. **Soil:** Soil on which the fruit tree is grown affects the time of maturity.
    - Grapes are harvested earlier on light sandy soils than on heavy clays.
3. **Size of planting material**
    - In pineapple, the number of days taken from flowering to fruit maturity is more by planting through large suckers and slips than by smaller ones.
4. **Closer spacing:** Close spacing of hill bananas hastens the maturity.
5. **Pruning intensity:** It enhances the maturity of Flordasun and Sharbati Peaches.

6. **Girdling:** Process of constricting the periphery of a stem which blocks the downward translocation of CHO, hormones, etc. beyond the constriction which rather accumulates above it. In Grape vines it hastens maturity, reduces the green berries in unevenly maturity cultivar and lowers the number of short berries. It is ineffective when done close to harvest. CPA has an additive effect with girdling.

MATURITY INDEX

The factors for determining the harvesting of fruits and vegetables according to consumer's requirement, type of commodity, *etc* can be judged by visual means (colour, size, shape), physical means (firmness, softness), chemical analysis (sugar content, acid content), computation (heat units and bloom to harvest period), physiological method (respiration).

Importance of maturity indices

- Ensure sensory quality (colour, flavour, aroma and texture) and nutritional quality.
- Ensure an adequate post-harvest shelf life.
- Facilitate scheduling of harvest and packing operations.
- Facilitate better marketing and price.

These are indications by which the maturity is judged. Various indices are as follows

1. **Fruit Colour**: Fruit skin or flesh colour changes as the fruit matures or ripens. These changes can be determined subjectively by the harvester. However, colour meters and colour charts have been developed for determining harvest times for apples, tomatoes, peaches, chilli, peppers *etc.* However, some fruits do not exhibit any perceptible colour changes during maturation and thus this parameter cannot be effectively used. Colour changes also differ among different cultivars of the same fruit.

   For example, the Hayward cultivar of kiwi fruit maintains its green flesh during maturation, while, 'Sanuki Gold' cultivar changes gradually to golden yellow colour. Some cultivars of avocado also maintain their green skin colour during maturation.

2. **Size and shape**: Maturity of fruits can be assessed by their final shape and size at the time of harvest. Fruit shape may be used in some instances to decide maturity. For example, the fullness of cheeks adjacent to pedicel may be used as a guide to maturity of mango and some stone fruits.

3. **Firmness:** Some fruits may change in texture during maturation and these changes can be used to determine the harvest time. Textural changes are detected subjectively by touch or gentle squeezing. However, objective measurement can be achieved using pressure tester and texture-analyser.
4. **Soluble Solids Content and Starch content:** During maturation, starch in non-climacteric fruits is converted to sugars. In climacteric fruits, starch accumulates during maturation. Therefore, harvest maturity can be determined by measuring the sugar content or starch content. Usually, the sugar content is measured in terms of total soluble solids content using a Brix hydrometer or refractometer. Starch content is measured using iodine to qualitatively determine the amount of starch. This method is used in determining the maturity of apple and pear cultivars whereby the fruit is cut into two pieces and dipped into a solution containing potassium iodide and iodine.
5. **Seed development:** It can also be used as an index of fruit maturity, *e.g.* endocarp hardening for stone and fibre development for dessert in mango.
6. **Calendar date:** Maturity in perennial fruit crops grown in seasonal climate is more or less uniform from year to year, so calendar date for harvest is a reliable guide to the commercial maturity. Fruit set refers to the transition of a flower to fruit after fertilization. It usually involves rapid cell division and expansion of the ovary and development of seeds. In some fruits, the time taken between fruit set until the fruit starts showing signs of maturity has been recorded and this can be used to determine harvest time. Time of flowering is largely dependent on temperature and the variation in number of days from flowering to harvest can be calculated for some commodities by the degree days- concept. For example, in Alphanso and Pairi mango varieties, it takes about 110 to 125 days after fruit set for surface colour to change from dark green to olive-green and flesh colour from white to pale yellow, while, Langra and Mallik take 84 and 96 days after fruit set, respectively to attain harvest maturity.
7. **Heat units:** Harvest date of newly introduced fruits in a widely varying climate can be predicted with the help of heat units. The heat requirement for fruit growth and development of each cultivar can be calculated in terms of degree days. Maturity at higher temperature is faster as the heat requirement is met earlier. This heat unit helps in planning, planting, harvesting and factory programmes for crops such as peas, mango and tomato for processing.
8. **Specific gravity:** The specific gravity of fruit can be considered as an index for maturity. Water has a specific gravity of 1.00 and common salt

solution (2.5% NaCl) has a specific gravity of 1.02 and both are used in the maturity grading of mango fruits. *e.g.* specific gravity of mango ranges between 1.01 to 1.02.

Maturity indices for important fruit crops

1. Mango

- Change in fruit shape (fullness of the cheeks)
- Skin colour changes from dark green to light green or yellow or red (according to variety)
- Change in flesh colour from greenish-yellow to yellow to orange.
- One or two fruits fall from the tree naturally - *Tapka* stage
- Reduction of flesh firmness
- Specific gravity (1.0-1.02 for Alphonso & less than 1.0 for Dashehari).
- White powdery like appearance on skin of mature mango.
- Lenticels become more prominant
- TSS 12-15%
- Days to fruit set (110-125 days for Alphonso and Pairi).

2. Banana

- Disappearance of angularity in a cross section *i.e.* degree of fullness of the fingers.
- Skin and pulp ratio (1.20:1.40 for Dwarf Cavendish).
- Drying of plant parts (brittleness of floral ends).
- Acid content (0.25%)
- Starch index.
- Days to fruit set (90 days for Dwarf Cavendish).
- Bananas are harvested mature-green and ripened upon arrival at destination markets.

3. Citrus

- All citrus are non-climacteric fruit, meaning that they ripen gradually over weeks or months and are slow to abscise from the tree.
- External colour changes during ripening from green to yellow or orange in colour.

- Juice content (35-50%)
- TSS 12-14% for mandarin and 10-12% for sweet orange
- Acidity (mandarin: 0.4%, sweet orange: 0.3%)
- The best indices of maturity for citrus are internal: Brix (sugar), acid content and the Brix/acid ratio

4. Sapota

- Skin colour shows a dull orange or potato colour with a yellowish tinge when scratched with finger nails.
- The scurf content on the surface of the fruit will be minimum and easily fall off.
- The content of milky latex almost drops to zero.
- Weight of fruit 65-70 g
- Specific gravity 1.025-1.057.

5. Papaya

- Change of skin colour from dark-green to light-green with some yellow at the blossom end (colour break).
- Papayas are usually harvested at colour break to 1/4 yellow for distant market or export and at 1/2 to 3/4 yellow for local markets.
- The latex of the fruit becomes almost watery.
- TSS should be around 7-11%.

6. Guava

- Guava fruits are picked at the mature-green stage (colour changes from dark to light-green).
- Specific gravity 1.0.
- TSS 12-14 %
- Fruit generally takes about 17-20 weeks from fruit set to maturity.

7. Jackfruit

- Jackfruits can reach very large size (as much as 90 cm long, 50 cm wide, and 25 kg in weight), depending on the cultivar, production area and fruit load on the tree.
- A dull, hollow sound is produced when the fruit is tapped by the finger.

- Colour change from green to yellow to brown is used as an indication of maturity and ripeness stage.
- The last leaf of peduncle turns to yellow.
- Fruit spines become flattened and widely spaced.
- Development of a typical aromatic odour.

8. Pineapple

- For local market- when 25% of surface changes to yellow colour

  For long distance market – when all the eyes are still green and have no traces of yellow colour (75-80% maturity)
- Flattening of eyes with slight hollowness at the centre.
- Specific gravity 0.98-1.02
- A minimum soluble solids content of 12-14% and a maximum acidity of 1% will assure flavour acceptability by most of the consumers.

9. Annona

- Change in skin colour from dark-green to light-green or greenish-yellow.
- Days to full bloom (100-115 days).
- Other indicators include appearance of cream colour between segments on the skin.
- Widening of the gap between fruit carpals or segments.

10. Aonla

- Change of seed colour from creamy white to brown black.

11. Pomegranate

- Fruits are ready for harvest between 135-170 days after anthesis.
- External red or yellow colour (depending on cultivar)
- The persistent calyx at the interior end of the fruit curves inward and become hard and dry at the time of maturity
- Flattening of rind on the outer sides of fruit.

12. Ber

- Fruit matures 150-175 days after flowering.
- Changes colour from green to golden yellow colour.

- Attainment of full size of a particular cultivar with softening of pulp.
- Seed/stone ratio: 12 to 18.
- TSS 15-18%.

12. Date palm

It can be harvested at 3 different stages;

- Doka (*Khalal*) - Fruit become hard, yellow or pink or red, TSS- 30 to 45%, astringency present or absent depending upon the cultivar and edible stage.
- Dang (*Rutab*) - Softness stating at the tip of the fruit, tannins and astringency disappears, lose weight and moisture content is about 35-40% at edible stage.
- Pind (*Tamar*) - Fully ripe fruit, lose weight, TSS-60 to 84% at edible stage.

Maturity indices for important vegetables

1. Tomato

- **Immature green:** It is the stage of fruit achieved prior to full development of seeds and before surrounding of seeds by a jelly like substance. The fruits are harvested at this stage only for frying purpose.
- **Mature green:** It is the stage of fruit when it is fully grown, and show brownish ring at the stem scare on removal of calyx and light green colour at blossom end changes to yellowish green and seeds are surrounded by jelly substances filling seed cavity. The fruits at this stage are harvested for shipment to long distance and for long storage too.
- **Turning (breaker stage):** It is the stage of fruit when one-fourth of the surface at blossom end shows pink colour. The fruits at this stage are harvested for local markets.
- **Pink stage:** It is the stage of fruit when three-fourth of the fruit surface shows the pink colour. The fruit at this stage are also harvested for local markets.
- **Hard ripe (red ripe) stage:** It is the stage of fruit when nearly the whole fruit skin shows red or pink colour but flesh is still firm. The fruits at this stage are harvested for table purpose, processing and for the extraction of seeds too.
- **Over ripe stage:** It is the stage of fruit when the fruit is fully red coloured and soft. At this stage, the fruits can be used only for the extraction of seeds, not for table purpose and processing since the fruits onward start decaying.

2. Capsicum

- **Green pepper varieties:** Fully mature green fruits should be harvested before ripening.
- **Red and yellow varieties:** Fully mature green fruits should be harvested at the onset of colour change (33%) for distant market while for local market, 75-90% colour change should be considered.
- Pepper fruits at the time of harvest should be firm and crisp.

3. Onion

Bulbs are considered mature when the neck tissues begin to soften and tops are about to abscise and decolourizes. Maturity can be judged by the neck of the plants drying up, tops falling over while the leaves are still green.

4. Sweet Potato

When the leaves turn yellow and begin to shed, tubers can be harvested. The cut surface of immature tubers shows dark greenish colour while the colour will be milky white in fully mature tubers.

5. Okra

Immature green tender fruits should be picked 3rd to 5th day from the time of first pod formation or 3 to 7 days after flowering depending upon variety and purpose (local or export market). Okra should be harvested when the pods are bright green, fleshy and seeds within are small.

6. Moringa

Fruits of sufficient length and girth are harvested before they become fibrous.

7. Cucumber

Fruits can be harvested from 45 days after sowing. The tender fruits (for salad) can be harvested on 8th to 10th day of flowering.

8. Bottle Gourd

Fruits should be light green, 30-35 cm long, tender with little pubescence persisting on the skin.

9. Muskmelon

- Fruits are generally harvested 60-70 days after sowing, 30-40 days after anthesis and 25-30 days after setting, observing other changes of outer colour of the skin depending upon varieties.
- Muskmelon is generally picked at 'half-slip' stage for commercial marketing (part of the pedicle remains attached to the fruit, *i.e.* abscission layer is not fully developed). Sugar and flavor are not found optimum at this stage. Full slip is stage at which the pedicle separates easily from the fruit with little or no pulling. Fruits for distance market should be harvested when mature but before fully ripeness to minimize to breakdown in texture and damage during transport.

10. Watermelon

- The fruits are ready for consumption in about 30-40 days after anthesis
- The curly tendril closest to the point of fruit attachment is often shriveled or dried.
- The portion of fruit resting on ground starts turing colour from creamy white to yellow.
- On ripening, the rind become hard enough that resists penetration of thumbnail.
- The sugar content of fruit measured as soluble solids using hand refractometre reaches 10% or more in flesh near center of fruit.
- On thumbing, the immature fruits give out metallic ringing sound and the ripened dull hollow sound.

11. Garden pea

- Early cultivars require as few as 1000 heat units to achieve maturity, whereas late sown cultivars may require more than 1600 heat units.
- The pods are harvested when they are filled, tender, having high sugar content and showing colour change from dark green to light green. Any delay in harvesting turns the pods to poor quality due to conversion of sugar into starch, and this conversion takes place more rapidly at high temperature.

12. Chilli

- Chilli should be harvested at fully mature prior to the colour change from green to red while approaching the ripe stage.

13. Potato

- Yellowish and drying of haulms

14. Cabbage

- Solidity, firmness, squeaking of heads indicates maturity.

15. Cauliflower

- Curd size and colour are deciding factors. Snow white or creamy white, compact curds surrounded by turgid green leaves.

References

Girdhari Lal, Siddappa, G. S. and Tandon, G. L. 1998. Preservation of Fruits and Vegetables. ICAR, New Delhi.

John, P.J. 2008. A Handbook on Postharvest Management of Fruits and Vegetables. Baya publishing House, Delhi.

Pandey, P.H. 1998. Principles and Practice of Postharvest Technology. Kalyani Publisher, Ludhiana.

Srivastava, R.P. and Sanjeev Kumar, 1998. Fruit and Vegetable Preservation Principles Practices. Oxford and IBH Publishers, New Delhi.

Sudheer, K.P. and Indira, V. 2007. Postharvest Technology of Horticultural Crops. New India Publishing, Agency, New Delhi.

Thompson, A.K. 1996. Postharvest Technology of Fruits and Vegetables. Blackwell Science, London.

Wills, R.B.H., Glassan, W.B. Mc, Graham, D., Lee, T. H. and Hall, E.G. 1981. Postharvest– An Introduction to the Physiology and Handling of Fruits and Vegetables. Wills, R. H.H., Granda, London.

16

Precision Farming and Protected Cultivation: Concepts and Analysis

Sanjeev Kumar, A. K. Pandey, Hetal Rathod Dushyant D. Champaneri and S.N. Saravaiya

Introduction

- Agriculture with its allied activities is the largest sector in India with 70% dependency of its rural households for their livelihood and 82% of farmers in the country are small and marginal.
- Agriculture sector alone contributes about 14.5% towards gross domestic product (GDP) of India.
- Food grain production has experienced a considerable jump from about 51 million tonne during 1950 to 291.95 million tonne in 2019-20 as per second advance estimates. The production of horticulture products has doubled over the past quarter century and the value of global trade in horticulture crops now exceeds that of cereals. For the first time, horticulture production surpassed food production in India during 2013-14 and continues to excel currently.
- India is characteristically a country of small agricultural farms, where approximately 80% of total land holdings in the country are less than 2 ha (5 acres) with 30% irrigated land only.
- India produces wide variety of agricultural products because of its varied agro-climatic regions but low farm productivity is major concern, which is around 33% of the best agricultural farms world over. Indian farmers can get more remuneration from the same piece of land with fewer inputs, once the productivity gets increased.
- Asia and the Pacific region accounts for more than 70% of the global agricultural population but has only 30% of the world's arable land. An increase in yield of these regions over the years has largely been due to

excessive and indiscriminate use of inputs like irrigation, seeds, fertilizers, pesticides, agrochemicals *etc*. and at considerable expense of its natural resources.

- Soil erosion in this region due to water and wind exceeds the natural soil formation by 30-40 folds. The problem of water quality deterioration is also very serious. Pollution of drinking water in tobacco and rice ecosystems of Malaysia, and of groundwater near vegetable fields in Japan are just two examples. However, the rate of increase in population demands much higher rate of increase in food production, while maintaining harmony with the environment-the core concept of sustainable agriculture.
- Most of the pesticidal recommendations are also region specific, which are on a few random observations of pest density. Because of low quality of agrochemicals often due to adulteration and increased pest resistance, indiscriminate application is now common especially in high valued crops. Hefty subsidies on electricity for pumping water in countries like India have led to over exploitation of groundwater and misuse on agricultural farms leading to water logging and salinity.
- Comprehensive and reliable information on land use/cover, soils (extent of waste lands and degraded lands), agricultural crops, water resources (both surface and underground), natural hazards/ calamities like drought and flood and agro meteorology is essential. Season-wise information on crops, their acreage, vigour and production enables the country to adopt suitable measures to meet shortages if any and implement proper support and procurement policies.
- With increasing population, urbanization and over exploitation of natural resources, there has to be a paradigm shift in farmers' perception from production towards productivity and profitability. In current scenario, agriculture is facing major challenges due to shrinking land holding, depleting water and other related resources. There is an urgent need for adopting farmer friendly location-specific production and management strategies in a concerted manner to achieve vertical growth in horticulture production with ensured quality of produce and judicious use of natural resources for better return per unit of area. In this context, precision farming has potential of utilizing resources per unit of time and area efficiently for achieving targeted production in horticultural crops.
- Precision agriculture is a key component of the third wave of modern agricultural revolutions. The first agricultural revolution came with the advent of increased mechanization during the period from 1900 to 1930s. Each farmer produced enough food to feed about 26 people during this time. The

1960s prompted the "Green Revolution" with new methods of genetic modification, which led to each farmer feeding about 155 people. It is expected that the global population will reach about 9.6 billion by 2050 and food production must effectively double from current levels in order to feed every mouth. With new technological advances in agriculture through precision farming, each farmer will be able to feed 265 people on the same acreage.

Concept of Precision Farming

Precision Agriculture (PA) also known as precision farming, prescription farming, variable rate technology (VRT), site-specific farming (SSF), site-specific management (SSM), site-specific crop management (SSCM), is considered as the vibrant agricultural system of the 21st century, as it symbolizes a better balance between reliance on traditional knowledge as well as information and management intensive technologies.

Professor Pierre C. Robert, who is considered as the father of precision farming defined precision farming as precision agriculture is not just the injection of new technologies but it is rather an information revolution, made possible by new technologies that result in a higher level, a more precise arm management system. **It basically means adding the right amount of input at the right time and the right location within a field.**

Precision agriculture (PA) ("precision farming", "site-specific farming", "farming by the foot", "spatially variable crop production", "grid farming"), is an innovative conception of agricultural production based on information technologies in crop production.

Precision Agriculture is the application of technologies and principles to manage spatial (space-related/geographic) and temporal (time related) variability associated with all aspects of agricultural production for improving production and environmental quality.

Precision Agriculture can also be defined as a comprehensive system designed to optimize agricultural production through the application of crop information, advanced technology and management practices.

US House of Representatives, 1997 have given the following two definitions of precision agriculture

Precision Agriculture (PA) is integrated information and production based farming system that is designed to increase long term, site specific and whole farm production efficiency, productivity and profitability, while minimizing unintended impacts on wildlife and the environment.

Site-specific crop management (SSCM): It is a form of precision agriculture whereby decisions on resource application and agronomic practices are improved to fit well to the requirements of crop and soil based on heterogeneity of the field.

The United States Department of Agriculture (USDA) refers this kind of agriculture "as needed farming" and define it as "a management system that is information and technology based, site specific and uses one or more of different sources of data like soil types, crops, nutrients, pests, moisture or yield for optimum profitability, sustainability and protection of the environment".

In simpler words, precision farming can be defined as information and technology (IT) based farm management system to identify, analyze and manage variability within fields by doing all practices of crop production in right place at right time and in right way for optimum profitability, sustainability and protection of the land resource.

In India, the average land holdings are very small, even with large and progressive farmers. The more suitable definition for precision farming in the context of Indian farming scenario could be- precise application of agricultural inputs based on soil, weather and crop requirement to maximize sustainable productivity, quality and profitability *i.e.* minimum input-maximum output approach.

Rationale of Precision Farming

The "Green revolution" of 1960's has made our country self sufficient in food production attributable to High Yielding Varieties (HYVs), fertilization, irrigation, pesticides and increase in cropping intensity and lastly mechanization of agriculture. However, Green Revolution Technologies (GRTs) has also led to some challenges, which needs to be addressed for increasing productivity of crops and simultaneously safeguarding ecological interest.

1. Fatigue of Green Revolution

No doubt, India has experienced spectacular growth in agriculture, however the productivity of many major crops are still below the expected levels. India has not achieved even the lowest level of potential productivity of Indian high yielding varieties, while the world's highest productive country has crop yield levels significantly higher than the upper potential of Indian HYVs.

2. Natural Resource Degradation

The green revolution has also been associated with some negative factors leading to ecological imbalance. Indian environment statistics shows that about 182 million ha of the country's total geographical area of 328.7 million ha is affected by land degradation, of which large proportion of 141.33 million ha is due to water erosion, 11.50 million ha due to wind erosion, and 12.63 and 13.24 million ha are due to water logging and chemical deterioration (salinity and loss of nutrients), respectively. On the other hand, India shares 17% of world's population, 1% of gross world product, 4% of world carbon emission, 3.6% of CO_2 emission intensity and 2% of world forest area. Hence, there is a dire need to convert this green revolution into an evergreen revolution with farming systems approach that can help to produce more from the available land, water and labour resources without ecological and social imbalance.

Traditional farming	Precision farming
Arable site is considered as a homogenous for applying treatment at field level.	Arable site is regarded as different from one point to another and considered as heterogeneous at field level.
Nutrient management is based on representative samples taken from a field.	Nutrient management is based on point like (site specific) samplings taken at specific sites and GPS.
Plant protection applications for diseases and insect-pests are generally made based on average survey done for plant damage.	Plant protection applications are made based on GPS and point (specific) like plant survey.
Selection and planting of plant species and variety is limited to overall arable site.	Plant species and variety are selected as per the suitability to a specific site/location.
Farm operations are adhered to same machinery in the arable site.	Farm operations by machinery are adjusted specifically to the arable site.
Plant stocks are minimally organized into homogeneous blocks at arable sites hence scope for better management is less.	Unified plant stocks are organized into homogeneous blocks at arable sites for better management.
Few data are available influencing decision making at arable site.	A lot of data are available to make conclusive decisions for better growth and development of a crop.

The critical differences between traditional and precision faming clearly define the prospects of precision farming for sustainable development of agriculture/ horticulture in India.

Objectives of Precision Farming

Higher profitability and sustainability: Maximum profit can be achieved in each zone or site of a field by optimizing precise application of inputs like variety, seed, fertilizer, herbicide, pesticide *etc.* as per crop demand, which can be determined by weather, soil characteristics (nutrient availability, texture and drainage) and historic crop performance.

Optimizing production efficiency: Identification of variability in yield potential may offer possibilities to optimize production quantity at each site or within each zone using differential approaches under given set of field conditions.

Optimizing product quality: Optimization of product quality by using sensors for the detection of quality attributes of a crop, which helps in taking decision on input application as per the target.

Increasing efficiency of inputs: Efficient use of inputs like fertilizer, seed *etc.* according to the yield potential of soil in a given location.

Effective and efficient pest management: To minimize the cost of inputs for crop production is one of the important objectives of precision farming, which can result in getting higher return and better environmental services. Site-specific variable rate application recommends the application of chemicals *i.e.* herbicides, pesticides at the area of problem with targeted approach in comparison to conventional farming methods.

Conservation of energy, water and soil: A comprehensive approach under precision farming begins from crop planning, assessing field variability, which includes those tillage practices which conserve the soil or disturb the soil to its minimum level. In addition to this, water is efficiently applied by using techniques like drip irrigation *etc.* with the principle of more crop per drop. In all such precision applications, very less energy is used thus leading to conservation of energy also.

Protection of surface and ground water: Safeguarding the environment by way of efficient use of inputs like fertilizers, chemicals etc. which prevents their leaching through ground water or as runoff.

Minimizing environmental impact: Better management decisions under precision farming are made to modify inputs for meeting out the production needs which ensures no or negligible loss of any applied input to the environment.

Minimizing risk: Risk management is a common strategy being adopted today by most of the farmers, which can be anticipated as income and environmental. In a production system, farmers often practice risk management by misjudging on the side of extra inputs, while the unit cost of a particular input is adjudged 'low'. Thus, a farmer may put an extra spray on, add extra fertilizer, buy more machinery or hire extra labour to ensure that the produce is produced/ harvested/ sold on time thereby guaranteeing a return.

Prospectives of Precision Farming

The effectiveness of precision farming is highly dependent on two factors: 1) Existence and assessment of variability within the fields/farms 2) Ability of a producer/farmer to identify and put into use the best management practices for each field's sub-area based on variability assessment. If the spatial data provided by precision farming is properly used, the following potential benefits can be realized:

Increased profits through increased efficiency: A producer can use agricultural inputs more efficiently as per the variability present for those components in specific sites of a field by applying more or less of them depending upon the specific requirements.

Reduced agronomic inputs: Producers/farmers can reduce their overall use of agricultural inputs like fertilizers, lime, chemicals etc. by making adjustment to a field's fertility level based on assessment, thus the amount of inputs applied to a specific site can specifically be applied.

Better record keeping: A large amounts of data on spatial records of inputs and outputs for agricultural fields generated by precision farming can help to develop more accurate management plans and strategies.

Improved production decisions: Precision farming can effectively be used to make decisions about land use. Profit maps of a farm showing the spatial distribution of a field's profitability can help a producer to make decisions about adoption of cropping systems at the best.

On-farm research: The ability to quantify the spatial performance of crops allows conducting more comparison trials within fields of a location. For instance, it is relatively simple to compare the performance of different varieties on different soil types, when practicing precision farming.

Reduced environmental impact: Production inputs are applied and used more efficiently under precision farming, so very low or minimum fraction of such inputs will leave the field through surface water and ground water. Reduced environmental impact will certainly have impact on the acceptance of farm produce in near future.

Property advantages: Many landlords are giving preference to farmers who can create yield maps and other files of spatial data for the fields under production practice. This spatial history may also increase the value of cropland.

More area farmed: The records generated under precision farming can allow managing more cropland effectively than the effectiveness observed in past.

Elements/Components of Precision Farming

A. Information or data base

Soil: Texture, Structure, Physical Condition, Moisture, Nutrients *etc.*

Crop: Plant Population, Plant Tissue Nutrient Analysis Status, Crop Stress, Weed patches (weed type and intensity), Insect or fungal infestation (species and intensity), Crop Yield, Harvest Swath Width *etc.*

Climate: Temperature, humidity, rainfall, solar radiation, wind velocity *etc.*

Field variability (spatially or temporally), soil related properties, crop characteristics, weed and insect-pest population and harvest data are important data bases for realizing the potential of precision farming.

B. Technology

Precision farming deals primarily with understanding the variability of field and its management over time and space. Precision farming includes a variety of tools of hardware, software and equipments like Computer system, Global Positioning System (GPS), Differential Global Positioning System (DGPS), Geographic information systems (GIS), Remote sensing, Sensors, Variable Rate Applicator, yield mapping technology/yield monitors etc. for variability assessment and its management in the field.

1. Computer System

Precision farming requires the acquisition, management, analysis and output of large amount of spatial and temporal data generated for a specific field. Computer software has now become better with time and the knowledge needed for managing variability and decision making on the field can be generated via such software to execute management strategies.

2. Global Positioning System (GPS)

GPS makes available continuous position (location) information in real time, while in motion. Positioning system works with the help of different constellations of satellites and developments of such different positioning systems are the main technological interventions bringing precision farming concept into reality. GPS receivers can directly be carried to the field or mounted on implements that allow users to return to specific locations to sample or treat such areas and helps to allow mapping of soil and crop measurements at any time. GPS provides accurate positioning system for field implementation of variable rate technology. It helps to manage application of inputs by equipments and identify the precise location of farm equipment within specific positions of the field for precision application of fertilizers and pesticides to the soil plant characteristics. GPS receiver with electronic yield monitors is generally used to collect yield data across the land in precise way. Global positioning systems (GPS) are widely used for mapping yields (GPS + combine yield monitor), variable rate planting (GPS + variable rate planting system), variable rate lime and fertilizer application (GPS + variable rate controller), field mapping for records and insurance purposes (GPS + mapping software) and parallel swathing (GPS + navigation tool).

3. Differential Global Positioning System (DGPS)

Differential Global Positioning System is a technique to improve GPS accuracy that uses pseudo range errors measured at a known location to improve the measurements made by other GPS receivers within the same geographic area. DGPS makes use of a series of satellites that identify the location of farm equipment within inches of an actual site in the field. The knowledge of a precise location helps to compare locations of soil samples and the laboratory results in a soil map, thus enable to prescribe fertilizers and pesticides to fit well to soil properties like clay and organic matter content and soil conditions (relief and drainage). Various tillage adjustments can be made with the help of information generated though this system. The system also allows monitoring and recording yield data as one goes across the field.

4. Geographic Information System (GIS)

Geographic information systems (GIS) are computer hardware and software with featured attributes and location data to produce maps. GIS includes organized collection of computer hardware, software, geographic data and personal designed to capture, store, update, manipulate, analyze and display all forms of geographically referenced information efficiently. Agricultural GIS stores series of information such as yield, soil survey maps, remotely sensed data, crop scouting reports and soil nutrient levels. GIS contains base maps like topography, soil type, nutrient level, soil moisture, pH, fertility, weed and pest intensity and can also integrate all types of information and interface with other decision support tools for application of recommended rates of nutrients or pesticides.

5. Remote Sensing

Remote sensing refers to the collection of data from a distance with the help of sensors, which can simply be hand held devices, mounted on aircraft or satellite. Remotely sensed data provide a tool for evaluating crop health and plant stress in relation to moisture, nutrients, compaction of soil, crop diseases and other plant health concerns oftenly and easily detected in overhead images. Remote sensing can reveal and analyze seasonal variability affecting crop yield and can be helpful to make timely management decisions to improve profitability for the current crop. Remotely sensed data obtained either by aircraft or satellite containing electromagnetic emittance and reflectance data of a crop can provide information useful for soil condition, plant growth, weed infestation *etc*. This type of information is cost effective and can be very useful for site-specific crop management programmes.

6. Sensors

Remote sensors are generally categorized as aerial or satellite sensors that can provide instant maps of field characteristics. An aerial photograph is optical sensors that can show variations in field colour corresponding to the changes in soil type, crop development, field boundaries, roads, water *etc*. Both aerial and satellite imagery can be processed to provide vegetative indices reflecting plant health also. Sensors are used to determine crop stress, soil properties, pest incidence *etc*. These can also be used to measure soil and crop properties as the tractor passes over the field, as a scout goes over the field on foot or as an airplane or satellite photographs of the field from the sky. Yield monitors are the primary sensing system that makes measurements. Commercial sensing systems are available that claim to measure soil properties and make changes in application rate.

Sensors can also be carried by a scout to the field and used to sense the health of plants and soil properties. These use light reflectance on the leaf to determine chlorophyll levels, which directly correlate nitrogen levels in the plant proportionally to the chlorophyll production. Several soil chemistry mini kits sensors are also available to measure soil pH, nitrogen, potassium, phosphorus *etc*. directly in the field.

7. Variable Rate Applicator

The variable rate applicator has three components namely computer controller, locator and actuator. A computer system mounted on a variable rate applicator uses the application map and a GPS receiver to direct a product delivery system that changes the amount and/or kind of product according to the application map *e.g.* Combine harvesters with yield monitors. Two methods are generally used for variable rate applications *viz*., first one is map based on historical data (previous or present year) that help the process control technologies to draw information from the GIS (prescription maps) and adjust fertilizer application, seeding and pesticide selection and application rate, thus providing proper management of the inputs. The second method uses sensors that can adjust the applications rates of equipments by detecting some characteristics of the crop or soil.

8. Yield Mapping Technology

Yield is ultimate indicator of variations of different agronomic/horticultural parameters in different parts within a field. So, mapping of yield and its interpretation, and correlation of that map with the spatial and temporal variability of different crop parameters help in developing management strategy for successive season of a crop. Present yield monitors measure the volume or mass flow rate to generate periodical record of quantity of harvested crop for that period and time periodic yield data is then synchronized with location obtained from onboard GPS system to create most common colour coded thematic map. Yield mapping can be carried out easily in mechanized crops, for example, loading cells weigh the crop passing on a conveying belt or an array of sonic beam mounted over the grape discharge chute to estimate the volume and the tonnage of fruit harvested. Radio-frequency identification (RFID) or bar code tags on the bins have also been used for yield mapping in which a weighing machine is combined with a tag reader and a GPS to record the weight and the place of each bin. The data collected are used to produce yield maps of the orchard.

Applications of Precision Farming

1. **Water Management:** Drip irrigation, sprinkler irrigation and fertigation are the important applications in precision farming for water management.
   a) **Drip Irrigation**: It is a regulated or slow application of irrigation water through emitters at frequent interval around crop rhizosphere over a period of time. It saves water by 50-70%, reduces labour and energy cost and has almost negligible weeds problem.
   b) **Sprinkler Irrigation:** Water is sprayed into the air and allowed to fall on the ground surface somewhat resembling to rainfall. The spray is developed by the flow of water pressure through small nozzles.
   c) **Fertigation**: Application of fertilizers with irrigation water to the crop, which provide nutrients directly to the active root zone, less likely to be lost thus minimizing the wastage of expensive nutrients and ultimately helps in improving productivity and quality of produce.
2. **Surface Covered Cultivation**
   a) **Mulching:** It is used to reduce or increase the temperature, suppress weed growth and conserve soil moisture.
   b) **Soil Solarization:** It is a method of heating the soil by covering it with transparent polythene sheets during hot periods to control soil-borne pathogens.
3. **Controlled Environmental Condition**: It includes greenhouse, polyhouse, grow tunnels, shade net houses *etc*., which are framed structures covered with a transparent or translucent material providing protection and creating congenial environment for growth and development of crops.
4. **Organic Farming:** It uses natural and biodegradable inputs which deliberately avoids the use of synthetic fertilizers/ or chemicals *etc*. It mainly

includes the use of vermicompost, manures, bio-fertilizers, animal husbandry, green manures, biological management and crop rotation *etc*.

5. **Precise Space Utilization**

   a) **High Density Planting (HDP):** HDP is a system of planting more number of plants than the optimum or normal through manipulation of tree architecture. Different methods like genetically dwarf scion cultivars, dwarfing rootstock and interstock, training and pruning, use of growth retardants are generally adopted to manage the size of plants to fit well under HDP.

   b) **Meadow Orcharding:** It is also known as ultra high density planting system which accommodates 20000-100000 plants/ha which is literally called as grass land. In order to maintain tree form, severe top pruning is practiced similar to mowing of grassland.

6. **Micro Propagation**: It refers to the production of plant from very small plant parts, tissues or cells grown aseptically in a test tube or containers under controlled environment.

7. **Integrated Pest Management (IPM)**: A pest management strategy which involves different components such as use of resistant varieties, managing the natural predators, adopting cultural practices, physical, mechanical, biological, curative methods and judicious application of pesticides to minimize pest population to a state below threshold level thereby avoiding economic injury to plants.

8. **Integrated Nutrient Management (INM):** It involves application of nutrients from organic as well as inorganic sources with the aim of improving physio-chemical and biological properties of the soil.

Limitations of Precision Farming under Indian Conditions

- Small sized farms/ small land holdings.
- Presence of heterogeneous cropping systems.
- Precision farming technology demands high initial capital investment.
- High cost of obtaining and analyzing site specific data.
- Complexity of precision tools and techniques requiring new technical skills.
- Precision farming as a new technology needs to be narrated and demonstrated to farmers with its on-farm and off-farm impacts on yields for better adoption.
- Lack of local technical expertise.

- Uncertainty on returns from investments on new equipments and information management system.
- Culture and attitude including reluctance of farmers towards adoption of new technologies and lack of awareness about environmental problems.
- Knowledge and technological gaps.

Protected Cultivation

Introduction

Water requirement is expected to rise upto the level of 1093 billion cubic meters (BCM) and 1447 BCM by 2025 and 2050, respectively, of which 910 and 1072 BCM constituting 83.26 and 74.08%, respectively of water would alone be required for irrigation purposes. Various applications of precision farming like water management, surface covered cultivation, controlled environment structure, organic farming, precise space utilization, micro-propagation/tissue culture, integrated pest & disease management, integrated and site-specific nutrient management, controlled environmental condition holds a great promise in present day situation. Amongst these, controlled environment structure is gaining momentum due to the efforts being made by Indian Government at various levels. The environmental constraints prevailing in a region can be overcome by adopting appropriate structures and environment control measures, thus almost round the year cultivation can be made with increased productivity by 25-100 % and in certain cases even more, as well as conservation of irrigation water by 25-50%. Protected farming offers itself as alternate farming method with much higher carrying capacity.

Protected cultivation is a technology which employs the principle of greenhouse effect with the involvement of structural engineering component to design location-specific protected structures for creating optimum/ near to optimum growing conditions and technology approaches for proper growth and development of plants inside such structures. The cladding/covering materials in greenhouse selectively allow solar radiations of short wavelengths to pass through it, which falling on ground are radiated back as long wavelength radiation and trapped as thermal energy by cladding material inside the greenhouse. This phenomenon is known as "*Greenhouse Effect*". Therefore, protected cultivation can be defined as the technology which uses structural engineering aspects to design protected structures meant to regulate inside environmental factors (temperature, relative humidity, light, CO_2) partially or fully along with some specific production techniques for better crop growth and development. A greenhouse refers to a framed structure covered with transparent and translucent materials supported by galvanized iron, bamboo or/ any other wooden material, wherein an enclosed area is used for crop cultivation under partially or fully controlled conditions. Greenhouse technology is the most intensive form of commercial cultivation and could well be called as **"Food Factories"**.

Potential benefits of protected cultivation

- Most suitable for cultivating high valued/off-season crops.
- Provides favourable micro-climate conditions for growth and development of plants.
- Higher yield with better quality per unit area.
- Conserves soil moisture thus needs less irrigation.
- Better management of pests and diseases.
- Helps in hardening of tissue cultured plants.
- Helps in raising early/ off-season nurseries.
- Round the year propagation of planting materials.
- Protects plants from wind, snow, rain, birds, hail *etc.*
- Generates self-employment opportunities for young India.
- Quality seed production.

Outlook of protected cultivation in India

- Apart from the pioneer work on research and development of protected cultivation of vegetables in India by Defence Research and Development Organization (DRDO) during 1960s at its one of the establishments namely, Field Research Laboratory, Leh now known as Defence Institute of High Altitude Research (DIHAR), the use of greenhouse technology started only during 1980s. It was mainly practiced as research activity initially, might be due to our focus was mainly on achieving self-sufficiency in food grain production during that period of time.
- Indo-American Hybrid Seeds India Pvt. Ltd., Bangalore was the pioneer to make use of greenhouse technology for commercial productions of flower seeds in India since 1965.
- The National committee on the use of Plastics in Agriculture (NCPA) was constituted in March, 1981 by Government of India under the Ministry of Petroleum Chemicals and Fertilizers. NCPA started to work under the Department of Agriculture & Cooperation with effect from 1993. The committee was reconstituted in 1996 in order to make it more effective and to focus its endeavour in a coordinated manner for promoting the plasticulture applications in Horticulture. The committee was reconstituted as National

Committee on Plasticulture Applications in Horticulture (NCPAH) in 2001 with Head Quarter at New Delhi.

- **Precision Farming Development Centres (PFDCs)** were also established in State Agricultural Universities (SAUs), ICAR Institutes such as IARI, New Delhi, CIAE, Bhopal and CISH, Lucknow and IIT, Kharagpur in order to promote precision farming and plasticulture applications for high-tech horticulture. These centers have been operating as hub-centers of plasticulture and precision farming in respective states. NCPAH established five new Precision Farming Development Centres at Bhopal, Imphal, Leh, Ludhiana and Ranchi during 2008-09 under the centrally sponsored scheme Micro Irrigation. (https://www.ncpahindia.com/).
- Centre for Protected Cultivation Technology (CPCT) was established at IARI, New Delhi in the year 1998-99 as demonstration farm and commissioned in January 2000 as Indo-Israel project undertaken jointly by the Government of India through Department of Agricultural Research & Education (DARE), ICAR and the Government of the State of Israel through the Centre of International Cooperation (MASHAV) and Center for International Agricultural Development Cooperation (CINADCO). The project was established to demonstrate peri-urban, high technology methods of growing flowers, vegetables and fruits. Collaboration with Israel Government concluded in 2004 and the unit was re-designed as Centre for Protected Cultivation technology (CPCT).
- National Horticulture Board (NHB) was set up by Government of India in April 1984 on the basis of recommendations of the "Group on Perishable Agricultural Commodities", headed by Prof. M. S. Swaminathan, the then Member (Agriculture), Planning Commission, Government of India. NHB started to provide financial support in components like open field cultivation, **protected cultivation** and primary processing since 1995-96 through soft loans till 2000. Thereafter, NHB launched its credit linked subsidy schemes under different components including protected cultivation.
- National Horticulture Mission (NHM) was launched in 2005-06 as a result of which significant progress has been made in horticultural crops through various interventions including protected cultivation.
- Rashtriya Krishi Vikas Yojana was initiated in 2007 as an umbrella scheme for ensuring holistic development of agriculture and allied sectors by allowing states to choose their own agriculture and allied sector development activities as per the district/state agriculture plan (DAP/SAP). Based on feedback received from States, the experiences garnered during implementation in the 12th FY Plan (2012–2017) and inputs provided by stakeholders, RKVY

guidelines have been revamped during November, 2017 for a period of 3 years i.e. 2017-18 to 2019-20 as **RKVY- RAFTAAR (Remunerative Approaches for Agriculture and Allied Sector Rejuvenation)** to enhance efficiency, efficacy and inclusiveness of the programme. Creation of value resources and protected cultivation (Green House/ Poly House/ Shade Net House infrastructures) is one of the components of infrastructure development under Horticulture Sector.

- **Mission for Integrated Development of Horticulture (MIDH)** is a Centrally Sponsored Scheme being implemented since 2014-15 for holistic growth of the horticulture sector covering fruits, vegetables, root and tuber crops, mushrooms, spices, flowers, aromatic plants, coconut, cashew, cocoa. MIDH subsumed ongoing missions/schemes of the Ministry- National Horticulture Mission (NHM), Horticulture Mission for North East & Himalayan States (HMNEH), National Horticulture Board (NHB), Coconut Development Board (CDB) and Central Institute for Horticulture (CIH). All States including North Eastern States and UTs are covered under MIDH. **An area of 2.31 lakh ha has been covered cumulatively in India under protected cultivation during the period from 2005-06 to 2017-18 under MIDH**.
- Horticulture Mission for North East and Himalayan States (**HMNEH**) was launched by Government of India in order to improve livelihood opportunities and bring prosperity to the North Eastern Region (NER) including Sikkim in 2001-02. Considering the potential of Horticulture for socio-economic development of Jammu & Kashmir, Himachal Pradesh and Uttarakhand, Technology Mission was extended to these States during 2003-04. HMNEH is based on the "End-to-End Approach" taking into account the entire gamut of horticulture development with all backward and forward linkages in a holistic manner. Government of India has approved the implementation of Technology Mission for Integrated Development of Horticulture in North Eastern States including Sikkim, Jammu & Kashmir, Himachal Pradesh and Uttarakhand (**TMNE**) during 11th FY Plan. As per the approval of GOI, the scheme will herein after be implemented in the name of HMNEH.
- The All India Coordinated Research Project on PET (formerly known as AICRP on Application of Plastic in Agriculture) become operational in 1988 during 7th FY Plan to undertake research and extension activities pertaining to water management, **protected farming**, post-harvest management *etc*. In 12th FY Plan, the project became operative at fourteen Centres located in different agro-ecological regions with its coordinating unit located at Central Institute of Post-Harvest Engineering and Technology (CIPHET), Ludhiana.

Skill India Initiative and Protected Cultivation

Skill India is a campaign launched by Government of India on 15th July 2015 with an objective of training over 40 crore people in India in different skills by 2022. It includes various initiatives of the government like "National Skill Development Mission", "National Policy for Skill Development and Entrepreneurship", "*Pradhan Mantri Kaushal Vikas Yojana* (PMKVY)" and "Skill Loan Scheme".

National Skill Development Mission

The National Skill Development Mission was officially launched by the Government on the occasion of World Youth Skills Day to create convergence across sectors and States in terms of skill training activities. Further, to achieve the vision of 'Skilled India', the National Skill Development Mission would not only consolidate and coordinate skilling efforts, but also expedite decision making across sectors to achieve skilling at scale with speed and standards. It is being implemented through a streamlined institutional mechanism driven by Ministry of Skill Development and Entrepreneurship (MSDE). The Mission is supported by three other institutions- National Skill Development Agency (NSDA), National Skill Development Corporation (NSDC) and Directorate General of Training (DGT) – all of which have horizontal linkages with Mission Directorate to facilitate smooth functioning of the national institutional mechanism. Seven sub-missions namely (i) Institutional Training (ii) Infrastructure (iii) Convergence (iv) Trainers (v) Overseas Employment (vi) Sustainable Livelihoods (vii) Leveraging Public Infrastructure have been proposed initially to act as building blocks for achieving overall objectives of the Mission.

Pradhan Mantri Kaushal Vikas Yojana (PMKVY)

Pradhan Mantri Kaushal Vikas Yojana (PMKVY) is a flagship scheme of MSDE meant to enable a large number of Indian youth to take up industry relevant skill training helping them in securing a better livelihood. Individuals with prior learning experience or skills are assessed and certified under Recognition of Prior Learning (RPL). Under this scheme, training and assessment fees are completely paid by the Government. Key components of the scheme include short term training, recognition of prior learning, special projects, *kaushal* and *rozgar mela*, placement and monitoring guidelines.

National Skill Development Corporation (NSDC)

NSDC was also setup as a one of its kind, Public Private Partnership Company with the primary mandate of catalyzing the skills landscape in India.

Agriculture Skill Council of India (ASCI)

To enable the creation and sustainability of support systems required for skill development, the Industry led Sector Skill Councils have been established by MSDE. Amongst these sector skills, **ASCI** is exclusively working towards capacity building by bridging gaps and upgrading skills of farmers, wage workers, self-employed and extension workers engaged in organized/unorganized segments of agriculture and allied Sectors. With the development of 169 Qualification Packs, ASCI has covered the various segments like Farm Mechanization and Precision Farming, Agri-Information Management, Dairy Farm Management, Poultry Farm Management, Fisheries, Animal Husbandry, Post-Harvest Supply Chain Management, Forestry & Agro Forestry, Watershed Management, Amenity Horticulture & Landscaping, Production Horticulture, Seeds Industry, Soil Health Management, Commodity Management, Agri Entrepreneurship & Rural Enterprises *etc*.

ASCI in its endeavour to develop/upgrade the skills of Indian agriculture and its allied workforce is collaborating with all key stakeholders including 32 Agriculture/ Horticulture, 17 Veterinary/Animal Sciences/ Fisheries Universities, 8 CSIR Institutes, 1 Skill University and Open University in order to align the short term skill training programmes with ASCI.

Basic considerations for protected cultivation

The design of greenhouse should be based upon sound scientific principles, which facilitates better plant growth under controlled or semi-controlled environment. There are some aspects which need to be given due consideration while starting protected cultivation.

Site Selection

While selecting the site for greenhouse, the following points should be considered for optimum growth and development of plants:

- The site should be free from shadow.
- The site should be at a higher level than the surrounding land with adequate drainage facility.
- Availability of good quality irrigation water and electricity to run the fan & pad cooling system.
- pH of the irrigation water should be in the range of 5.5 to 7.0 and EC between 0.1 to 0.3 $mScm^{-1}$.
- pH of the soil should be in the range of 5.5 to 6.5 and EC between 0.5 to 0.7 $mScm^{-1}$.

- Proximity to motorable road to take advantage of market for inputs supply and sale of prodiceprocedure.
- Soil needs to be sterilized after every 3 to 4 years to avoid built up of soil-borne pathogens particularly nematodes or one may resort to soil-less cultivation.

Orientation

Accurate orientation of protected structure can provide congenial environmental conditions for growth and development of crops grown inside the greenhouse. Following points should be taken into consideration while deciding the orientation of a greenhouse depending upon light intensity, and direction and velocity of wind. The pictorial depiction of greenhouse orientation is shown in Fig. 1.

- Orientation of the single span greenhouse should be directed towards East-West (length to width), while it should be in North-South direction for multi-span type greenhouses (gutter direction).
- Slope along the gutter should not be more than 2%, while it should not be more than 1.25% along the gable side.
- Ventilators should open on the leeward side in naturally ventilated greenhouse.
- Single span greenhouse should have long axis perpendicular to the direction of wind to protect it from wind damage.
- Wind breaks should be placed at least 30 meters away on North-West side of the greenhouse.
- There is less availability of solar radiation and the sky remains cloudy during rainy season in tropical regions, hence so effective ventilation is very important to manage temperature and humidity by keeping proper orientation of polyhouse.

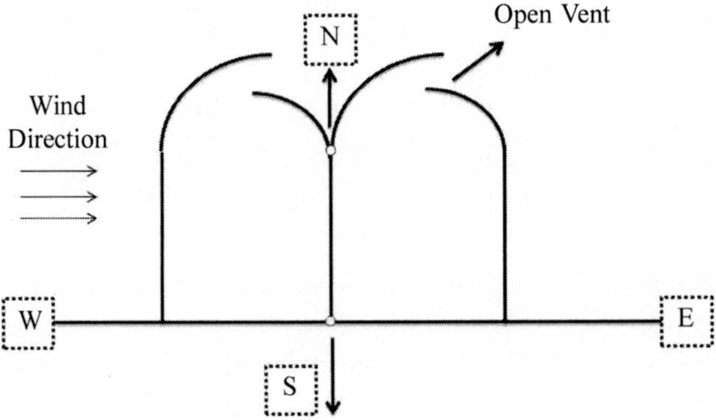

Fig. 1. Greenhouse Orientation

Design considerations for greenhouse

The structural engineering aspect of protected structure has is to be dealt with utmost care, so that it must have very good strength to carry the following loads.

a) **Dead load**: Weight of all permanent construction, cladding, heating and cooling system, water pipes and all fixed service equipments to the frame.

b) **Live load**: Weights laid over by use of components like hanging baskets, shelves and persons working on top of the structure. The greenhouse has to be designed to carry live load maximum of 15 kg m^{-2}. Each member of roof should be capable of supporting 45 kg of concentrated load when applied at its centre.

c) **Wind load**: The structure should be able to withstand wind speed of 150 km hr^{-1} and at least 50 kgm^{-2} of wind pressure.

d) **Snow load**: It is to be taken as per the average snowfall of the location. The greenhouse should be able to take dead load plus live load or dead load plus wind load plus half the live load.

The greenhouses are to be fabricated of Galvanized Iron (GI) Pipes. The foundation can be 60 cm x 60 cm x 60 cm or 30 cm diameter and one meter depth in plain cement concrete (PCC) of 1:4:8 (one part of cement, four parts of sand and eight parts of coarse aggregate) ratio. The vertical poles should also be covered to the height of 60 cm by PCC with a thickness of 5cm, which will avoid the rusting of the poles.

Size

The size of the greenhouse needs to be selected based on availability of the land and the cost, which may vary depending upon the types of structure specifically fabricated for the region. Depending upon the market access and experience of greenhouse cultivation, it is suggested to start with a naturally ventilated greenhouse of minimum size as it would require less initial capital investment along with operational expenditure. However, experienced farmers/ entrepreneurs may decide to go for larger greenhouse depending on their scale of operation and project costs.

Height

Height is one of the most important aspects of greenhouse design. The height of the structure directly impacts natural ventilation, stability of the internal environment and crop management. The ideal central height and side/gutter height of naturally ventilated greenhouse as well as for greenhouse with fan & pad system is tabulated below:

Component	Type of structure		
	NVPH Small (up to 250 m²)	**NVPH large (> 250 m²)**	**Greenhouse with Fan & Pad system**
Central height	3.5 -4.5 m	5.5-6.5 m	<5.5 m
Side/ gutter height	2.5-3.0 m	4.5-5.0 m	As per NVPH size

Both types of greenhouse can be made in single or multi-span structures. A multi-span greenhouse can be constructed for an area more than 200 m^2 and is economical in terms of construction material and required control/ monitoring equipment.

Cladding Materials

Cladding (Covering/glazing) materials are the most important constituent of protected structures as they have direct influence on microclimate inside the greenhouse. Transmission for global radiation (Ultra Violet-UV and Photosynthetically Active Radiation-PAR), heat radiation (Near Infra Red-NIR and Far Infra Red-FIR), insulating effect, sensitivity to ageing (mainly UV degradation), permeability for humidity (water), mechanical strength (tensile and impact), fire behaviour, investment costs, available dimensions etc. are the important properties, which must be taken into consideration while evaluating cladding materials for greenhouse. So, there are different types of cladding materials, which can be used in greenhouses depending upon the suitability to the region and project cost.

Glass as cladding material

Glass is preferred by many growers as covering material for greenhouses considering greenhouse technology a long term investment. Greenhouse glass is available in several types like float, tempered, laminated and frosted or hammered glass, typically the most common being the float glass. Float glass is a transparent glass with a high light admittance to ensure sufficient light in the greenhouse. These days, greenhouses are often fitted with safety glass like tempered or laminated. Tempered glass is 4 to 6 times more resistant to shattering than float glass and it splits into small square pieces making it unlikely to cause injury even upon breakage. Glass when used as cladding material has advantages like 1) Have relatively long life span. 2) Have superior light transmitting properties and less excessive relative humidity problems. 3) Glass “breathes” (the glass laps between panes allow air to enter), while polyethylene, acrylic, and polycarbonate structured sheet greenhouses are airtight making excessive humidity in the growing area and undesirable water drip on the plants.

Flexible plastic films as cladding material

Flexible plastic films namely polyethylene, ethylene tetrafluoroethylene (Tefzel), polyvinyl chloride (PVC) and polyester are used as covering material in polyhouses. **Polyethylene** has always been and still is the primarily preferred cladding material for greenhouses in most parts of the world because of its low cost compared to other cladding materials. Tefzel T2 film is another addition in plastic coverings of greenhouses. A single layered Tefzel T2 film has light transmission of 95%, which is greater than that of any other covering material, while a double layer has a light transmission of 90 %.

Polyvinyl chloride has a very good tolerance to wear and tear and very little effect of oxidation, thus making it desirable material for greenhouse covering. However, heat and light may break down PVC film in 2 to 3 years. PVC film has characteristic property of reducing transmission of infrared rays, thus there are less chances of heat loss during night in greenhouses covered with PVC film than that of polyethylene.

Polyester film is costlier than that of polyethylene film, but it is well known for its durability and longer life expectancy. It has high advantage of high light transmittance equivalent to glass and provides freedom from static electrical charges, thereby avoids settlement of dust over the sheet.

As per the Indian Standard (IS 15827: 2009), the plastic films should have following important characteristics:

Important characteristics of covering materials

S.No.	Type of plastic films	Characteristics	Uses
1.	Normal film	a) Transparency more than 80% b) Low Greenhouse Effect	Forcing and Semi-Forcing crops
2.	Thermic clear film	a) Good Transparency b) High infrared (IR) Effectiveness	As normal film,when greater infrared(IR) effectiveness is desired.
3.	Thermic diffusion	a) Diffusion Light b) High infrared (IR) Effectiveness	As normal film, whengreater infrared (IR)effectiveness and Diffusing light is desirable.

Rigid plastic as cladding material

Fiberglass-reinforced plastic (FRP) rigid panel, polycarbonates, and acrylics are used as rigid plastic coverings. Generally, rigid plastics have very good light transmission, but it usually decreases over a period of time because of aging and yellowing of coverings due to the impact of UV radiations.

FRP rigid panel has been in use as greenhouse coverings since the 1950s but its popularity has declined in recent years. Fiberglass is available in flat and corrugated configurations. Corrugated panels are commonly used for greenhouse roofs, as its corrugated shape lends strength and rigidity to the panels. Flat panels are usually used for sidewalls, windows, and vents. Although FRP panels are classified as a rigid plastic, but are flexible enough to be bent in a curve to fit the framework of greenhouse structures namely quonset type or arch type.

Polycarbonate is one of the most widely used structured sheet material in greenhouses today. Polycarbonate has higher percentage of direct radiation versus diffused radiation in comparison to polyethylene film. It is similar to acrylic film in heat retention properties and allows about 90% of light transmission of light. It is considerably stronger and impact resistant, more flexible, and only flammable when an active flame is maintained in contact with the material. Although, the initial cost of polycarbonate is high, but its life expectancy is 10 to 15 years. Polycarbonate has high impact strength of about 200 times that of glass.

Acrylic has been used for many years and is considered to be the most suitable rigid transparent plastic for greenhouse covering. Acrylics have excellent clarity, light transmission, high impact resistance and are flame retardant and UV stabilized, and has a textured surface for better diffusion of light thereby preventing condensation drip.

Agro shade nets as cladding material

Each type of cultivated plant needs a specific shade for different diversed phases of its development. Agro shade nets fulfill the requirement of giving appropriate micro-climate conditions to the plants. These nets are designed to protect the crops and plants from UV radiation and simultaneously provide protection against climate conditions like temperature variations, intensive rain, winds *etc*. Better growth conditions can be achieved for a crop due to the semi-controlled or fully controlled micro-climate conditions created in the growing area with shade nets for getting higher yield. All nets are UV stabilized to fulfill expected lifetime at the area of exposure and have high tear resistance, low weight for easy and quick installation. A wide range of agro shade nets are available in different shading percentages or shade factor *i.e.* 15%, 35%, 40%, 50% 75% and 90%, which are defined on the basis of percentage of shade delivered to plants growing under them (for example 35% shade factor means the net will cut 35% of light intensity and would allow only 65% of light intensity to pass through the net).

Comparison of different kinds of covering materials

Type	Durability(year)	Transmission		Maintenance
		Light (%)	Heat (%)	
Polyethylene	01	90	70	Very high
Polyethylene UV resistant	02	90	70	High
Fibre glass	07	90	05	Low
Tedlar coated Fibre Glass	15	90	05	Low
Double strength glass	50	90	05	Low
Poly carbonate	50	90	05	Very Low

It is important to distinguish between diffuse and direct radiation in respect of Photosynthetically active radiation (PAR) radiation. Direct radiation refers to the radiation, which reaches the earth directly from the sun without being reflected. However, most of the sun radiation is scattered by small water droplets in the atmosphere on cloudy days and reaches the earth as diffuse radiation. When the radiation is direct there are big differences between the locations in the greenhouse with and without shadow (40% compared to 85%). Generally, transmittance for diffused light is 10% lower than the transmittance for direct light but it reaches the lower parts of high plants better, which is better for overall growth of plants. The transmittances for diffused and direct light of some cladding materials are given for the wavelength band of 400-700 nm (PAR) in Table 1.

Table 1. Transmittance for direct (perpendicular) and diffused light of some important cladding materials of greenhouses (PAR, 400-700 nm).

Material	Thickness	Transmittance for direct, perpendicular light	Transmittance fordiffuse light
Glass	4 mm	89-91%	82%
PE (Polyethylene) film	200 μm	89-91%	81%
EVA (Ethylenevinylacetate) film	180 μm	90-91%	82%
PVC (Polyvinylchloride) film	200 μm	87-91%	-
ETFE (Ethylene tetra fluoroethylene) film	100 μm	93-95%	88%
PVDF (Polyvinylidene fluoride) film	100 μm	93-94%	85%
PC (Polycarbonate) sheet	12 mm	80%	61%
PMMA (Poly Methylmethacrylate) sheet	16 mm	89%	76%
PC (Polycarbonate) zigzag sheet (double)	25 mm	88%	79%

Criteria for selecting greenhouse structure

Galvanized iron (GI) pipe based greenhouse must conform to ISI Trade mark with minimum 2 mm wall thickness

The structure should have following features

- Must have stub type-anchoring foundation, galvanized and nut bolted structure having options for future expansion.
- Should be strong enough to withstand various types of load such as live load, dead load, crop load, wind and snow load.
- G.I. gutter should be in a single piece having width of 500 mm and thickness of 1 mm.
- Must have aerodynamic shape along all peripheries to resist high wind velocity.
- The structure must be designed in such a way that it can resist wind velocity up to 150 $Kmhr^{-1}$.
- There must be provision for fixing of UV stabilized plastic film with aluminum/ G.I. profiles & zigzag spring lock.
- Should be strong enough to support the load of internal service systems and plant foliage.
- UV stabilized plastic film of 200 micron thickness conforming to Indian Standard IS 15827: 2009 should be used.
- Should be easy to service, cover and recover with cladding material.
- Should be easy to operate and must be cost effective.

SWOT Analysis of Protected Cultivation

It is a powerful tool to plan a strategic process consisting of internal and external forces. To develop a plan that takes into consideration many different internal and external factors and maximizes the potential of **Strengths** and **Opportunities** while minimize the impact of the **Weaknesses** and **Threats**.

Strengths (S)

- Maintain, build and pull
- Internal Positive Factors- may be capitalized
- High input use efficiency.
- Favourable micro-climate for better crop growth.
- High yield, off-season production and quality produce with export potential.
- Innovative, integrated, knowledge-oriented technology for sustainable horticulture.
- Production of high quality planting material/ off-season nurseries.

- Good Governance.
- Subsidy of 50% being imparted by Govt. of India and more than that by State Governments.
- Positive attitude of progressive farmers on new technology (s).
- Good co-ordination and linkage with SAUs of India, other Agricultural Institutes and allied departments.
- Big niche markets in various states of India as well as easy access to metropolises and other big cities.
- Well developed infrastructure and co-operatives.
- Higher income than outdoor production.

Weaknesses (W)

- Prioritize, optimize, remedy or exit
- Internal Negative Factors- may be improved
- Non-availability of indigenously bred/developed varieties/hybrids exclusively for protected cultivation.
- Lack of cost effective and location specific design for greenhouses.
- High initial and operation cost of protected structure.
- Non-availability of tools and implements facilitating crop production operations under protected structures.
- Knowledge and technology gaps in greenhouse management are prominent.
- Lack of marketing intelligence and strategy.
- Limited possibilities to demonstrate innovative products.
- Shortage of product specialist, sales agents/distributors, maintenance companies.
- Environmental concern over the disposal of non-degradable plastic films used in protected structures

Opportunities (O)

- Prioritize, Optimize
- External Positive Factors- mat take advantage.
- Standardization of designs of protected structures for different agro-climatic regions of the country.
- Standardization of designs for vertical farms in ambitious smart cities of India

- Arrival of new technology (s) for farming community.
- Development of POPs, INM and IPDM modules for greenhouses crops.
- Organic farming for selected class of mass.
- Revival of youths' declined interest in agriculture through hi-tech agribusiness and generation of young-skilled India.
- Removal of trade barriers.
- Promotion and extension provided by government sector.
- Increasing horticultural product for domestics and export demand.
- Tremendous scope for establishment of processing units.
- An unfulfilled customer need.
- Scope to generate marketable surplus and also surplus for export.
- Inclusion of modern concept of protected cultivation (Hydroponics, Aeroponics, Bioponics etc. in ambitious Smart Cities of India)

Threats (T)

- Counter
- Internal Negative Factors- may be prevented and reduced
- Very little or negligible work on development of techno-economic feasible and region specific cropping scheme.
- Market infrastructure may be inadequate keeping in view the increasing product flow.
- Uncertainty about market stability and farmers do not get remunerative price.
- Exploitation by middleman in the market chain.
- Pest and diseases are main problem when the greenhouse structures grow older.
- Lack of standards for crop production in protected culture.
- Lack of standards for post harvest management of the produce.
- Inefficiency of farm management makes farmers difficult to get the advantages of globalization and trade liberalization.
- Cash flow and bad debts problems.
- Unawareness on new regulations.

Impact of NHM and HMNEH on Protected Cultivation in India

Protected cultivation happens to be the first priority sector considering total expenditure made in the 12th Five Year Plan and the highest total number of sample beneficiaries next to area expansion component as revealed in a study conducted by Institute for Social and Economic Change, Bengaluru on impact of NHM and HMNEH for **protected cultivation**. The outcomes of the study are mentioned below:

- Out of the 4033 sample beneficiaries, 715 beneficiaries (17.7%) have taken up protected cultivation and the majority of structures adopted by beneficiaries are polyhouse (61.25%) followed by net house (16%) and mulching (15.24%).
- Total crop-wise area under protected cultivation was worked out to 2916 m^2 (0.2 acres) per beneficiary.
- The responses of beneficiaries indicated that the materials used by the farmers and the material supplied by system integrators are of good quality.
- A majority of these structures are being used for the cultivation of vegetables and flowers. About 42% of the sample beneficiaries have acquired their planting materials or seeds from private nurseries and 32% from the public nurseries.
- The productivity of vegetables and fruits worked out to be 9 kg per m^2 and 3 kg per m^2 at national level. Among states, Haryana, Assam, Bihar, Madhya Pradesh, J & K, Gujarat, Meghalaya and Himachal Pradesh account for a productivity of more than 20 kg of vegetables per m^2. With respect to flower production, Assam was found to have highest productivity (249 cut flowers per m^2) followed by Maharashtra (240 cut flowers per m^2).
- The average annual net income derived by farmers from protected cultivation amounts to Rs. 70.5 per m^2 at all India level. The income derived was highest in respect of cut flowers and vegetables to the tune of Rs. 147 per m^2 and Rs. 71 per m^2, respectively. Among states, Kerala and Madhya Pradesh have earned more than Rs. 600 per m^2 in respect of cut flowers and more than Rs. 300 per m^2 by Kerala and J & K for vegetable crops. In Haryana, the net income earned from fruits amounts to Rs. 721 per m^2.
- A large proportion of the respondents indicated that the productivity level under protected cultivation has increased significantly as against open cultivation.

Following recommendations were emerged out from the impact assessment study, which will help to improve the effectiveness of protected cultivation:

- Farmers must be encouraged to follow crop rotation for disease management and sustaining the yield levels.
- Flower growers should go for rooting chambers inside polyhouse or outside. The provision of subsidy for this component has to be considered which, in turn, will help to reduce the cost of planting materials.
- The polyhouse intervention has to be integrated with drip irrigation, rainwater harvesting, pack house and farm pond which are demanded by the farmers.
- Protected cultivation must be integrated with pollinators since pollinators mediate in increasing the yield.
- The government is currently giving subsidy to the extent of 50% of the cost of green house at national level. The farmers are unable to mobilize the rest of the amount. Hence, farmers may be given 75% as subsidy so as to encourage higher adoption of protected cultivation.
- Several beneficiaries are depending on middle men for lack of access to markets and market information. In order to get better prices, more rural markets and connectivity to other markets may be established.
- There is a need for providing a minimal assistance for maintenance of protected cultivation structures as repair and maintenance are quite high.
- New concepts in protected cultivation such as vertical farming, hydroponics, aquaponics and growing vegetables under controlled lighting conditions should be taken into consideration. MIDH may take up the matter with ICAR to undertake R & D for developing separate protocols and SOPs.
- Collaboration may be initiated with certain private organizations having advanced infrastructure and technologies and with countries like Israel, Denmark, Germany, Holland, Australia and Japan for specialized protected cultivation technologies.

Technological challenges

1. Non-availability of indigenously bred/ developed varieties/ hybrids for protected cultivation
2. High initial and operation cost of protected structures.
3. Lack of work on development/ standardization of designs of protected structures for different agro-climatic regions of the country
4. Very little work on the development of POPs, INM and IPDM modules for greenhouse crops.
5. Non-availability of tools and implements for facilitating crop production operations under greenhouse. For instance electric vibrator for pollination in greenhouse tomatoes.

6. Very little or negligible work on development of techno-economic feasible and region specific cropping scheme.

Strategies to strengthen protected cultivation

1. Establishment of National Research Centre on Protected Cultivation.
2. Development of suitable varieties/ hybrids for protected cultivation
3. Identification of new and potential crop types for protected cultivation
4. Development of cost-effective and location-specific designs of greenhouses
5. Development and use of new generation biodegradable polymers
6. Designing and simulation of photovoltaic greenhouse (PVG) systems for Indian conditions
7. Adoption of novel technologies in cladding material like NIR reflecting, Cool plastic film, photo-selective film that creates cooler temperature in the greenhouse
8. Use of antivirus film i.e. UV block film that completely blocks UV radiation to adversely affect thrips and whitefly
9. Development of complete POP for greenhouse crops in different agro-climatic regions.
10. Identification of region specific and techno-economic feasible cropping sequences
11. Development of effective INM, IPDM and GAP modules for protected cultivation.
12. Designing and development of tools, devices and equipment for easy operations in greenhouse crops
13. Standardization of new age technologies like Hydroponics, Aeroponics, Nutrient Film Technique, Agro-voltaic systems, vertical farms *etc*.
14. Developing professional and skilled manpower
15. Use of solar panels in greenhouse
16. A mission on protected cultivation and vertical farms is needed in the country
17. Pollination management for crops under protected structures
18. A mission on protected cultivation and vertical farms is needed in the country.
19. Pollination management for crops under protected structures.

References

Abdullahi, H.S., Mahieddine, F. and Sheriff, R.E. 2015. Technology impact on agricultural productivity: A review of precision agriculture using unmanned aerial vehicles. Doi: 10.1007/978-3-319-25479-1_29.

Alka Singh and Sanjeev Kumar (Eds.), 2018. Training Manual on Skill Development in Protected Cultivation. 35/2018-19. 270p.

Anil Kumar Singh. Precision Farming. http://apps.iasri.res.in/ebook/EBADAT/6-Other%20Useful%20Techniques/14-Precision%20Farming%20Lecture.pdf

Anil Kumar, Tiwari, G.N., Subodh Kumar and Mukesh Pandey, 2006. Role of greenhouse technology in agricultural engineering. International Journal of Agricultural Research, 1(4): 364-372.

Anonymous, 2011. Compendium of Environment Statistics (http:mospi.nic.in/compendium-environment-statistics).

Baidya, A.S. 2017. Precision farming: A new vista for Indian horticulture. Rashtriya Krishi, 12 (2): 59-63.

Bradley, K. and Khosla, R. 2003. The Role of precision agriculture in cropping systems. Journal of Crop Production, 9 (1-2): 361-381. Doi: 10.1300/J144v09n01_02.

Brahma Singh, 2019. Precision Farming and Protected Cultivation. New India Publishing Agency, New Delhi.

Castilla, N. 2013. Greenhouse Technology and Management, 2nd Edition. CABI, Nosworthy, Wallingford Oxfordshire OX10 8DE, UK. CABI, Way 38 Chauncey Street, Suite 1002, Boston, MA 02111, USA.

Das, U., Pathak, P., Meena, M.K. and Mallikarjun, N. 2018. Precision farming a promising technology in horticulture: A Review. International Journal of Pure and Applied Bioscience. 6 (1): 1596-1606.

FAO, 2013. Good Agricultural Practices for greenhouse vegetable crops. FAO Plant Production and Protection Paper 217. Food and Agriculture Organization of the United Nations, Rome. www.fao.org/publications.

Hakkim, A.V.M., Joseph A.E., Gokul, A.J. and Mufeedha, K. 2016. Precision Farming: The future of Indian agriculture. Journal of Applied Biology and Biotechnology, 4 (06): 68-72.

http://www.drishtiias.com/upsc-exam-gs-resources-Precision-farming1of914052018.

http://www.fao.org/india/fao-in-india/india-at-a-glance/en/.

http://www.fao.org/india/fao-in-india/india-at-a-glance/en/.

http://www.iari.res.in.

http://www.ncpahindia.com/application.php?appl_code=9.

https://aicrp.icar.gov.in/pet/.

https://asci-india.com.

https://midh.gov.in.

https://web.archive.org/web/20150829000757.

https://www.msde.gov.in.

https://www.ncpahindia.com/precision-farming.

https://www.prsindia.org/report-summaries/economic-survey-2018-19.

https://www.researchgate.net/publication/283641594_Technology_Impact_on_Agricultural_Productivity_A_Review_of_Precision_Agriculture_Using_Unmanned _Aerial_Vehicles?enrichId=rgreq-cf6f2080dd4 7fc4bff95e8a3 75d4d8ac-XXX&enrichSource=Y 292ZXJQYWdlOzI4MzY 0MTU5NDtBUzoyOTQyNDg 5NDg3NDgyOTNAMTQ0NzE2NTgy NTEyMg%3D%3D&el=1_x_2&_esc= publicationCoverPdf

Iyengar, K.S., Gahrotra, A. Mishra, A., Kaushal, K.K. and Dutt, M. 2011. Greenhouse- A reference manual. National Comittee on Plasticulture Applications in Horticulture (NCPAH), NCPAH/TB/2010-11/13.

Liaghat, S. and Balasundram, S.K. 2010. A Review: The role of remote sensing in precision agriculture. American Journal of Agricultural and Biological Sciences, 5 (1): 50-55.

Mandal, D. and Ghosh, S.K. 2000. Precision farming -The emerging concept of agriculture for today and tomorrow. Current Science, 79 (12): 1644-1647.

Mandal, S.K. and Maity, A. 2013. Precision Farming for small agricultural farm: Indian scenario. American Journal of Experimental Agriculture, 3 (1): 200-217.

Manjunatha, A.V., Ramappa, K.B., Maruthi, I. and Parmod Kumar, 2017. Impact evaluation of National Horticulture Mission (NHM) and Horticulture Mission for North East and Himalayan States (HMNEH), ADRT Centre, Institute for Social and Economic Change, Bengaluru, Karnataka, 254p.

McMahon, R.W. 1992. An Introduction to Greenhouse Production. Ohio Agricultural Education Curriculum Materials Service. The Ohio State University, Columbus, Ohio.

Mishra, A., Sundaramoorthi, K., Chidambara R.P. and Balaji, D. 2003. Operationalization of Precision Farming in India. In: Map India Conference 2003. © GISdevelopment.net.

Mishra, J. and Paul, J.C. 2008. Greenhouse for rural development. In: Rural Infrastructure: Sanitation, Housing, Health Care (Verma, S. B., Singh, S.G.P. and Singh, S.K., Eds.), Sarup and Sons, New Delhi.

Model Bankable Project on Protected Cultivation in Haryana (www.nhm.nic.in).

Mondal, P. and Basu, M. 2009. Adoption of precision agriculture technologies in India and in some developing countries: scope, present status and strategies. Progress in Natural Science, 19: 659-666. Doi:10.1016/j.pnsc.2008.07.020.

NAAS, 2010. Protected Agriculture in North-West Himalayas. Policy Paper No. 47 (Rajendra Prasad and Pathak, P. S. Eds.), National Academy of Agricultural Sciences, New Delhi, pp 16.

Naved Sabir and Balraj Singh, 2013. Protected cultivation of vegetables in global arena: A review. Indian Journal of Agricultural Sciences, 83 (2): 123-135.

Patil, S.S. and Bhalerao, S.A. 2013. Precision farming: The most scientific and modern approach to sustainable agriculture. International Research Journal of Science and Engineering, 1 (2): 21-30.

Rains, G.C. and Thomas, D.L. Precision Farming- An Introduction. University of Georgia Research and Extension Personnel. https://www.researchgate.net/publication/277791290

Reddy, P.P. 2015. Greenhouse Technology. © Springer Science+Business Media Singapore. Sustainable Crop Protection under Protected Cultivation. Doi: 10.1007/978-981-287-952-3_2.

Rocha, J.V. Precision Farming and Geographic Systems. https://www.researchgate.net/publication/37680195_GIS_and_remote_sensing_application_on_precision_agriculture?

Sahoo, R.N. 2003. Precision farming: Concept and approaches. In: Proceeding of DOS funded 12th Winter School on Remote Sensing Application in Agriculture with special emphasis on Precision Farming, 2003: 1-10.

Sanjeev Kumar; S.N. Saravaiya, S.N. and Pandey, A.K. 2021. Precision Farming and Protected Cultivation: Concepts and Applications. Jaya Publishing House, Delhi.

Singh, A., Dhankhar, S.S. and Dashiya, K.K. 2015. Protected cultivation of horticultural crops. Doi: 10.13140/RG.2.218172.95364

Srinivasan, A. Relevance of Precision Farming Technologies to Sustainable Agriculture in Asia and the Pacific (http://www.sristi.org/mtsa/s_7_4.htm)

Stafford, J.V. 2000. Implementing Precision Agriculture in the 21st Century. Journal of Agricultural Engineering Research, 76: 267-275.

Sudduth, K.A. Engineering Technologies for Precision Farming. https://www.semanticscholar.org/paper/ENGINEERING-TECHNOLOGIES-FOR-PRECISION-FARMING-Sudduth/c81f739bf02 08f20a85bc58446c8119311b7d8cb#citing-papers.

Tran, D.V. and Nguyen, N.V. The concept and implementation of precision farming and rice integrated crop management systems for sustainable production in the twenty-first century.https://pdfs.semanticscholar.org/f67d/28db59c04514336f8b16860c7e25dd4e5e5f.pdf?_ga=2.228042840.454061046.1590062449-444517366.1512281757. http://agropedia.iitk.ac.in/content/precision-agriculture-glance.

Waaijenberg, D. 2006. Design, Construction and Maintenance of Greenhouse Structures. Proc. IS on Greenhouses, Environmental Controls and In-house Mechanization for Crop Production in the Tropics and Sub-tropics (Rezuwan Kamaruddin, Ibni Hajar Rukunuddin & Nor Raizan Abdul Hamid, Eds.). Acta Horticulturae, 710: 31-42.

Wagh, R.M. 2017. Precision farming, remote sensing, geographical information system: A new paradigm for agricultural production in India. International Archive of Applied Sciences and Technology, 8 (4): 04-09.

Wittwer, S.H. and Castilla, N. 1995. Protected cultivation of horticultural crops worldwide. HortTechnology, 5 (1): 6-23.

Zhang, C. and Kovacs, J.M. 2012. The application of small unmanned aerial systems for precision agriculture: A review. Precision Agric: An International Journal on Advances in Precision Agriculture, 13: 693-712.

17

Insect Pest Management in Fruit Crops

Snehal M. Patel, H.V. Pandya, V.P. Prajapati
P. R. Patel and Hemant Sharma

1. INSECT- PESTS OF MANGO

1. Mango hoppers: *Idioscopus niveosparus, I. clypealis, Amritodus atkinsoni* (Cicadellidae: Hemiptera)

Management

- Avoid close planting, as the incidence becomes very severe in overcrowded orchards.
- Orchards must be kept clean by ploughing and removal of weeds.
- Pruning of dense canopy to facilitate aeration and sunlight.
- Avoid excess use of nitrogenous fertilizers.
- Immediately after withdrawal of monsoon (September-October), off-seasonal trunk spraying with Malathion 50 EC @ 15 ml in 10 lit of water suppresses the adult hoppers from multiplying further and hinders their oviposition and also kills the newly hatched nymphs.
- Avoid application of all other synthetic pyrethroids as they have been reported to cause resurgence of hoppers.
- If overlapping generations of hoppers are found, spray imidacloprid @ 3 ml or acetamaprid 4 gm in 10 litres of water at panicle emergence stage.
- During fruit set stage, hoppers can effectively be controlled using imidacloprid @ 0.005% or thiamethoxam 0.0084%. Insecticides should be used only when hoppers reach to the Economic threshold level (5 nymphs & or adults per shoot during vegetative stage and per panicle during flowering stage).
- Spray dimethoate 30 EC, methyldemeton 25 EC or Malathion 50 EC @ 1.5 to 2.0 L in 1500-2000 L of water per ha or acephate 75 SP @ 1 g/L, phosalone 35 EC @ 1.5 ml/L, or new molecules like buprofezin 25 SC 1-2 ml/L of

water or imidacloprid 17.8 SL @ 2-4 ml/tree or lambda cyhalothrin 5 EC 0.5-1.0 ml/L of water (10 -15 L of water per tree).

- At low population pressure, spray 3 per cent neem oil or neem seed kernel powder extract @ 5%.

2. Stem borer: *Batocera rufomaculata* (Cerambycidae: Coleoptera)

Management

- Grow tolerant mango varieties *viz*., Neelam, Humayudin.
- Remove and destroy dead and severely affected branches with grubs and pupae.
- Avoid injury at the base of trunk while pruning.
- Beetles wherever found in the garden should be collected and destroyed.
- Remove alternative hosts like moringa, silk cotton in the near vicinity.
- The grubs should be extruded through hooked wires or destroy by injecting carbon disulfide and chloroform (2:1) with the help of injecting syringe or dilute kerosene or petrol @ 5 ml/hole or one celphos tablet (3 g aluminum phosphide) or apply carbofuran 3G 5 g per hole and plug with clay + copper oxychloride paste.
- Swab coal tar + Kerosene @ 1:2 (basal portion of the trunk at 3 feet height) after scraping the loose bark to prevent oviposition by adult beetles.

3. Fruit fly: *Bactrocera dorsalis* (Tephritidae: Diptera)

Management

- Sanitation in the orchard is most important.
- Collection and destruction of the fallen fruits.
- Racking the soil around the tree to expose and kill the pupae.
- Drench chlorpyriphos 20 EC @ 2.5 ml/L to the racked soil around tree trunk to destroy the pupa.
- Three weeks before the harvest, spray Deltamethrin 2.8 EC @ 0.5 ml/L + Azadirachtin (0.3%) 2 ml/L.
- If fruit fly is very serious (> 5/Surveillance Trap), give bait sprays on the tree trunks at weekly interval: (Bait spray is prepared by mixing 10g of Jaggery in one litre of water to which 2 ml of Deltamethrin (2.8 EC) is added).
- Methyl eugenol work as male sex attractant. Use Methyl eugenol traps *i.e.* 10 traps/hectare or 4-5 traps/acre.

- Grow the black *Tulsi* in or around the orchard and spray malathion 50% EC @ 10 ml/10 L on it at a fortnight interval to attract and destroy the male fruit fly. For effective management of this pest, cluster approach should be followed.

4. Mango nut weevil: *Sternochaetus mangiferae* (Curculionidae: Coleoptera)

Management

- Under-sized fruits left on the tree should be picked and destroyed.
- Undertake general cleanliness and destruction of the hibernating weevils on the bark during August. If the trees are few, bag the fruits with cloth or try paper bags for protection.
- Collect and destroy the fallen fruits and stones.
- Spray of malathion 50 EC @ 1 ml/L; Quinalphos @ 2 ml/L during Sept.-Oct. at marble stage of the fruits followed by second spray at 15 days interval.
- During non-flowering season, spray should be directed towards the base of the trunk.
- The infested bark should be washed with kerosene emulsion.
- Spray deltamethrin 1.5-2.0 L (1 ml/L of water) in 1500-2000 L water per ha after six weeks of fruit set.

5. Mango mealy bug: *Drosicha mangiferae* (Pseudococcidae: Hemiptera)

Management

- Remove weeds like *Clerodendrum inflortunatum* and grasses by ploughing during June-July.
- Plough orchards during summer to expose the eggs to natural enemies and extreme heat.
- Apply band of 20 cm wide alkalthene sheets (400 gauge) to tress in the middle of December (50 cm above the ground level and just below the junction of branching). Tie stem with jute thread and apply a little mud of fruit tree grease on the lower edge of the band.
- Release Australian ladybird beetle, *Cryptolaemus montrouzieri* @ 10/tree
- If necessary, spray dimethoate 30 EC or methyldemeton 25 EC or malathion 50 EC 1.5-2.0 L or chlorpryriphos 20 EC @ 3.0-4.0 L in 1500-2000 L water per ha.

- Once the pest reaches TO the top of the plant, control becomes rather difficult.

6. Bark eating caterpillar: *Indarbela tetraonis, I. quadrinotata* (Metarbelidae: Lepidoptera)

Management

- Kill the caterpillars by inserting an iron spike into the tunnels.
- Inject ethylene glycol and kerosene oil (1:3) into the tunnel by means of a syringe and then seal the opening of the tunnel with mud.
- Dip a small piece of cotton in any of the fumigants like chloroform or petrol or kerosene, introduce into the tunnel and seal the opening with clay or mud.

7. Flower gall midge: *Procystiphora mangiferae,* Erosomyia indica, Dasineura amaramanjarae (Cecidomyiidae: Diptera)

Management

- Spray dimethoate 30 EC or methyl demeton 25 EC @ 3.0-4.0 L in 1500-2000 L of water per ha (10-15 L of spray fluid per tree).
- Cutting of heavily infested plant parts at the early stage of infestation and destroying them.

8. Flower webber: *Eublemma versicolor* (Noctuidae: Lepidoptera)

Management

- Spray phosalone 35 EC @ 3.0-4.0 L in 1500-2000 L of water per ha (10-15 L of spray fluid per tree).

9. Mango leaf webber: *Orthaga exvinacea* (Noctuidae: Lepidoptera)

Management

- Remove and destroy the webbed leaves along with larvae and pupae. Conserve predators like carabid beetle *Parena lacticincta*, reduvid Oecama sp, parasitoid *Hormiusa* and fungus *Paecilomyces farinosus*.
- Three spray of 0.05 % quinalphos at fortnightly interval during June as well as October-December.

2. INSECT- PEST OF SAPOTA

1. Leaf webber or Chiku moth: *Nephopteryx eugraphella* Rongonot (Pyraustidae: Lepidoptera)

Management

- Remove and destroy all the infested dried clusters of leaf webs, buds and fruits to reduce infestation.
- Spray phosalone 35 EC 2 ml/lit or neem seed kernel extract 5% as and when new flush of flower buds set on trees.
- Two spray of nimbecidin 0.3 % (Neem based formulation) at 20 days interval.

2. Bud worm: *Anarsia achrasella* (Gelechiidae: Lepidoptera)

Management

- Spray phasalone 35 EC @ 3 L in 1500-2000 L water per ha.
- Remove and destroy all the infested leaves, buds and fruits to reduce infestation.
- Installation of traps containing black *tulsi* leaves extract along with dichlorvos during April to September.
- Install black *tulsi* leaves extract poison bait trap (8-10 traps/hectare) at 4.5 cm height from ground level of peripheral canopy of *chiku* tree followed by three sprays of profenophos 50% EC @ 10-20 ml in 10 L of water at 15 days interval starting from bud initiation. Change the traps at alternate days.
- Two spray of nimbecidin 0.3% (Neem based formulation) at 20 days intervals starting from March.

3. INSECT-PESTS OF BANANA

1. Rhizome weevil: *Cosmopolites sordidus* (Curculionidae: Coleoptera)

Management

- Use healthy and pest free suckers.
- Use healthy planting material and remove outer layer of rhizome and sundry for 3-4 days before planting after smearing with slurry of cow dung and ash.
- Adopt strict field sanitation by removing infected plants and destroying them.
- Trap the adult weevils by placing chopped pseudostem in the cropped area.
- Setting traps in the field using length-wise split pseudostem of 50 cm length. Adults attracted to it during nights may be collected and destroyed.

- Uproot and destroy infested rhizomes.
- Deep ploughing before planting to expose the weevils to sun and predators.
- Soil incorporation of carbofuran 3G @ 20 g/plant around pseudostem.
- Drenching with chlorpyriphos 0.1% emulsion in the soil before planting may provide some relief.

2. Pseudostem borer: *Odoiporus longicollis* (Curculionidae: Coleoptera)

Management

- Adopt good cultivation practices to improve weevil tolerance.
- Maintain healthy plantation by periodical removal of dry leaves and suckers.
- Field sanitation by removing and destroying the affected plants alongwith rhizome and also the destruction of pseudostem and rhizome of harvested plants is the most important method.
- Prune the side suckers every month.
- Do not dump infested materials into manure pit.
- Uproot infested trees, chop into pieces and burn.
- Application of carbofuran 3 G @ 30g/plant at planting and @ 15g/plant at 60th and 90th day after planting.
- Spray quinalphos 0.05% or chlorpyriphos 0.03% at the time of planting. In case of severe infestation, spray may be repeated after 3 weeks.

3. Thrips: *Helionothrips kadaliphilus*, *Thrips florum*, *Chaetanothrips signipennis* (Thripidae: Thysanoptera)

Management

- Spray methyl demeton 25 EC or dimethoate 30 EC @ 3.0-4.0 L in 1500-2000 L water per ha towards the crown and pseudostem base.

4. INSECT-PESTS OF GUAVA

1. Tea mosquito bug: *Helopeltis antonii* (Miridae: Hemiptera)

Management

- Undertake pruning to regulate proper penetration of sunlight inside the canopy.
- 3 sprays of phosalone 0.07% or dimethoate 0.05%, first at the time of flushing, second at early flowering and third at the time of fruit set.

2. Fruit fly: *Bactrocera diversus* (Tephritidae: Diptera)

Management

- Collect and destroy the damaged plant parts.
- Summer ploughing to expose and kill pupae.
- Harvest the fruits when slightly hard and green.
- Spray fenvalerate 20 EC @ 1 L or malathion 50 EC @ 2 L in 1500-2000 L of water per ha.
- Rake the soil around the tree to expose pupae.

3. Castor capsule borer: *Conogethes punctiferalis* (Pyraustidae: Lepidoptera)

Management

- Collect and destroy the damaged plant parts.
- Use light traps @ 1trap/ha to monitor the activity of adults.
- 2 Sprays of malathion 50 EC @ 3 L or dimethoate 30 EC @ 3 L in 1500-2000 L water per ha, first one at flower formation and next at fruit set.
- Destroy all the infested shoots, buds and fruits in the initial stage of infestation.
- In case of severe infestation, spray malathion 50 EC @ 10 ml/10 L.

4. Bark caterpillar: *Indarbela tetraonis* (Metarbelidae: Lepidoptera)

Management

- Kill the caterpillars mechanically by inserting an iron spike into the tunnels made by these caterpillars.
- Keep the orchard clean to prevent the infestation.
- Dip a small piece of cotton in any of the fumigants like chloroform or petrol or kerosene, introduce into the tunnel and seal the opening with clay or mud.
- In case of severe infestation, clean the affected portion of the trunk or main stem and insert into the hole a swab of cotton wool soaked in a 0.05 per cent emulsion of dichlorvos and seal the hole with mud paste for killing the larva in tunnels.

5. Scarlet Mite: *Brevipalpus phoenicus* (Tenuipalpidae: Acari)

Management

- Collect and destroy the damaged plant parts.
- Spray wettable sulphur @ 3 kg in 10 liter of water per ha.
- Prune regularly to reduce dense canopy

Minor pests

1. Aphids: *Aphis gossypii*
2. Guava Scale: *Chloropulivinaria psidii*
3. Whitefly : *Aleurotuberculatus psidii*
4. Thrips: *Selenothrips rubrocinctus*, *Porthesia scintillans*

5. PEST OF GRAPES

1. Stem girdler: *Sthenias grisator* Fabricius (Cerambycidae: Coleoptera)

Management

- Remove loose bark at the time of pruning to prevent egg laying.
- Cut the attacked branches below girdling point and burn the same.
- Hand collection and destruction of beetles.
- Spray Insecticides like quinalphos 25 EC @ 0.05% immediately after pruning and repeat it 2-3 times.

2. Flea beetle: *Scelodonta strigicollis* (Chrysomelidae: Coleoptera)

Management

- Remove the loose bark at the time of pruning to prevent egg laying.
- Shake vines to dislodge adult beetles into trays containing kerosene water and destroy them.
- Spray quinalphos 25 EC @ 0.05% after pruning.
- Remove the loose bark and spray with dichlorvos 0.075% in 1000 litres per hectare immediately after pruning and repeat 2-3 times after 10 days.

3. Thrips: *Rhipiphorothrips cruentatus* Hood (Thripidae: Thysanoptera).

Management

- Collect and destroy damaged leaves, fruits and flowers.
- Spray methyl o demeton 25 EC or dimethoate 30 EC @ 10 ml in 10 liter of water at 15 days interval.

4. Mealy bug: *Maconellicoccus hirsutus*, *Pseudococcus maritimus* (Pseudococcidae: Hemiptera)

Management

- Debark vines and branches.
- Use sticky traps on fruit bearing shoots at a length of 5 cm.
- Collect the damaged bark, leaves, twigs and stems along with mealy bug colonies and destroy.
- Use dichlorvos 76 WSC @ 0.15% in combination with fish oil rosin soap (25 g/L) as spray or dipping fruits for 2 min.
- Release exotic predator, *Cryptolaemus montrouzieri* @ 10 beetles/vine. 3 releases may be required depending on mealy bug population.
- Spray dimethoate 30 EC + kerosene oil (150 ml + 250 ml) in 100 ml of water or malathion 50 EC @ 0.1%
- Apply quinalphos in the soil at 25 kg/ ha to kill phoretic ants.
- Use of Predators: *Cryptolaemus montrouzieri*, *Scymnus craccivora, Mallada boninensis*, *Spalgis epius*, *Cacoxenus perspicax*, *Triommato craccidivora*
- Parasitoids: *Anagrus dactylopii, Allotropa sp., Gyanusoidea mirzai, Alamella flava.*

6. INSECT-PESTS OF BER

1. Ber fruit fly: *Carpomyia vesuviana* (Tephritidae: Diptera)

Management

- Collect and destroy the fallen infested fruits at alternate days.
- Cultivate fruit fly resistant varieties such as Safeda Ilaichi, Chinese, Sanaur-1, Mirchia, Tikadi and Umran.
- Collect and destroy fallen and infested fruits by dumping in a pit and covering with a thick layer of soil. Plough interspaces to expose pupae. Conserve parasitoids *Opius compensates* and *Spalangia philippinensis*.
- Use methyl eugenol lure trap (25/ha) to monitor and kill adults of fruit flies or prepare methyl eugenol and Malathion 50 EC mixture at 1:1 ratio and take 10 ml mixture/trap. Use bait spray combining molasses or jaggery 10 g/L and malathion 50 EC 2 ml/L or dimethoate 30 EC 1ml/L, two rounds at fortnight interval before ripening of the fruits.
- Spray Malathion 50 EC 1.0 L or dimethoate 30 EC 1.0 L or dichlorvos 700 ml at the time of flower formation and fruit set.

2. Ber fruit borer: *Meridarches scyrodes* (Carposinidae: Lepidoptera)

Management

- Collect and destroy the damaged fruits.
- Rake the soil periodically.
- Apply chlorpyriphos 1.5 D at 40 kg per ha around the trees prior to the attainment of marble stage of fruits.
- Spray Malathion 1.0 L or dimethoate 750 ml L at the time of fruit set, two rounds at 15 days/interval.

7. INSECT- PESTS OF COCONUT

1. Rhinoceros beetle: *Oryctes rhinoceros* (Scarabaeidae: Coleoptera)

Management

- Destroy and dispose all dead trees.
- Avoid manure pits in the vicinity of coconut gardens.
- Hook out beetles from crowns during peak period of infestation
- Rake and turn up the decaying manure to expose the developing grub, egg and pupae to sun drying and predation. Then apply the fungal culture of *Metarrhizium anisopliae* to manure pits during cooler months of October - December.
- Encourage reduviid predators, *Platymeris laevicollis.*
- In seedlings, place naphthalene balls @ 3 / tree, in the innermost three leaf axils once in 45 days.
- Soak castor cake @ 1 kg/5 lit of water in wide mouthed mud pots and keep them in the garden to attract and kill adults. Replace the slurry once in 30 days.
- Fermented toddy may be kept in wide mouthed earthern vessels in different places to attract the adults during night.
- The crown region may be properly cleaned during harvests and the adults may be hooked out using a long wire.
- Light traps may be set up to attract the adults during monsoon months and following rains during summer.
- The top-most three axils may be filled with a mixture of sand + Neem Seed Powder (2:1) once in three months (150 g/tree).
- Fill leaf axil with powdered marotti cake (Hydnocarpus) @ 250 g /palm during May, September and January as a prophylactic measure.

- Incorporate *Clerodendron infortunatum* whole plant in the breeding sites.
- Use aggregation pheromone traps Rhinolure @ 1/ha. Instal the trap at five feet from the ground level.

2. Red palm weevil: *Rhynchophorus ferrugineus* (Curculionidae: Coleoptera)

Management

- Remove and disposal of damaged and wilted trees.
- Avoid injuries on trunk and any injury should be plastered with clay or cemented with copper oxychloride.
- Avoid cutting green fronds.
- Root feeding with monocrotophos @ 10 ml + 10 ml water should be done after harvest of nuts. Observe a waiting period of 45 days.
- Set up attractant traps using mud pots with molasses / toddy 2.5 lit + acetic acid 5 ml + yeast 5 g + split tender coconut stems / petioles @ 30/ac.
- Insert 1-2 aluminium phosphide tablets inside the tunnel and plug all the holes with clay + copper oxychloride.
- Use aggregation pheromone trap @ 1/ha or use ferrolure in combination with food baits consisting of 1 kg sugarcane molasses + 5g yeast + 5 ml glacial acetic acid + split petioles of coconut taken in a bucket of 10 L capacity.

3. Black headed caterpillar: *Opisina arenosella* (Cryptophasidae: Lepidoptera)

Management

- Prompt detection of commencement of the infestation by surveillance and cutting and burning of infested fronds forma an important measure of control.
- Sprays of malathion, quinalphos or phosalone are found useful.
- In summer, release bethylids, braconid and eulophid parasitoids from January at 1:1:10 per tree.
- Root feeding with monocrotophos @ 10 ml + 10 ml water with a waiting period of 45 days after root feeding.
- Periodical release of *Persierola nephantidis* or *Bracon brevicornis* or *Elasmus nephanbdis* may be useful.

4. Coconut Eriophyid mite: *Aceria guerreronis Keifer (Acari : Eriophyidae)*

Management

i) Nutrients (per tree/year)

- Urea 1.3 kg, super 2.0 kg, potash 3.5 kg, neem cake 5 kg, borax 50 g, gypsum 1 kg, $MgSO_4$ 500 g, FYM 50 kg.

ii) Root feeding

- Root feeding with carbosulfan 15 ml + 15 ml water / tree at 45 days interval or fenpyroximate at 10 ml/tree

  Note: Pluck nuts before root feeding. After root feeding, next harvest should be done 45 days later.

iii) Spray

- Fenpyroximate 5 EC 1.0 ml/L of water.
- Triazophos 40 EC 5 ml/lit @ 2 ml / lit or carbosulfan 25 EC 2 ml/ lit in alternation with neem azal 1% 5ml/lit as spot application

5. White grub: *Leucopholis coneophora* (Melolonthidae: Coleoptera)

Management

- Summer ploughing exposes the immature stages.
- Raising of crop early in the Kharif season.
- Treat the seeds with chlorpyriphos @ 12 ml/kg of kernels.
- Apply carbofuran 3 G @ 30 kg per ha in the soil at or before sowing.
- Dust the soil with 10 kg chlordane 5 % per acre. Carry out this treatment during tilting and deep ploughing of soil at the time of pest abundance.

6. Slug caterpillar: *Parasa lepida, Contheyla rotunda* (Cochlidiidae: Lepidoptera)

Damage symptoms

- Defoliation, leaving only the midrib and veins

Management

- Remove the affected leaves in the lower rows.
- Collect and destroy the immature stages of the insects.
- Spray dichlorvos 76% EC at 2 ml/litre of water or carbaryl 50 WP 2g/l.
- Set up light trap to trap and monitor adults moth @ 5 traps/hactare.

References

Butani, D.K. 1984. Insects and Fruits. Periodical Expert Book Agency, New Delhi.

http://agspsrv34.agric.wa.gov.au/Ento/Surveillance/Mango%20leafhopper.html

http://t2.gstatic.com/images?q=tbn:ANd9GcRbK4UTUqEc2X4DXN5JTK9Svul0yITOtR1Ifcb0mX-YaJr_FOcFuQ

Reddy P.P. 2009. Advances in Integrated Pest and Diseases Management in Horticultural Crops (Volume 1: Fruit crops). Studium Press (India) Pvt. Ltd., New Delhi.

Reddy, P.P. 2010. Plant Protection in Horticulture (Vol. 1, 2 & 3). Scientific Publishers, Jodhpur.

Srivastava, R.P. 1997. Mango Insect Pest Management (first edition). International Book Distributing Co., Lucknow.

Verma. L.R., Verma A.K. and Gautam, D.C. (Eds.), 2004. Pest management in Horticultural Crops: Principles and Practices. Asiatech Publishers Inc., New Delhi.

18

Disease Management in Fruit Crops

V.P. Prajapati, P.R. Patel, Hemant Sharma, H.V. Pandya and Snehal M. Patel

1. DISEASES OF MANGO

Mango is considered to be the king of fruit. India is the largest producer and exporter of mango in the world. Mango possess unique nutritional and medicinal qualities apart from being a rich source of vitamins A & C, besides its attractive form and appearance, delicious taste and appetizing flavor, the ripe mango fruit according to nutritional experts is also highly invigorating, fattening, laxative and diuretic. Every part of mango from root to tip is used in a variety of ways. This crop is affected by many fungal, bacterial and other non parasitic diseases.

1. Mango Malformation Disease (MMD)

Mango Malformation Disease is a fungal disease of mango caused by several species of *Fusarium,* some yet to be described. Mango is the only known host of the disease. The disease spreads on a tree very slowly, but if left unchecked, can severely reduce yields. The main method of spreading MMD to new areas is through infected vegetative planting material. There is no evidence that the disease can spread on fruit or the seeds, or that it affects human health. It usually associated with the bud mite, *Aceria mangiferae* but the mites have been shown to spread the disease within a tree and not between the trees. Mango malformation, also known as bunchy top, is a very serious threat to the mango industry, particularly in northern India. The Etiology of the disease still remains obscure and diverse claims have been made about its causes, e.g., physiological, viral, fungal, acarological and nutritional.

Symptomatology

Three distinct types of Symptoms are produced.

1. Bunchy top of seedlings (BT)
2. Vegetative malformation (VM)
3. Floral malformation (MF)

Bunchy top of seedlings (BT)

Bunchy top phase (BT) appears on young plants in the nursery beds when they are 4-5 months old. Formation of a bunch of thickened small shootlets bearing small rudimentary leaves or occasionally several bunches arising from a leaf axil at the top or lower down the main shoot. These shoot lets are much thicker than main axis from which they arise. The shoot remains short and stunted. The growth of the plant is stopped and it gives an appearance of bunchy top.

Vegetative malformation (VM)

Induces short internodes forming bunches of various sizes. They are found at the top of the seedling and give a bunchy top appearance.

Floral malformation (MF)

And the secondary branches are transformed into vegetative buds and large number of small leaves and stems, which are characterized by appreciably reduced internodes and are compacted together giving a witches 'broom appearance. In other cases, the flower buds seldom open and remain dull green.

Causal Organism: *Fusarium moniliformae* var. *subglutinans.* Wollenw. & Reink.

Micro conidia are one or two-celled, oval to fusiform and produced from polyphialides. Macro conidia are rarely produced and are 2-3 celled and falcate. Asexual fruiting body of the fungus is sporodochium. Chlamydospores are not produced.

Mode of spread and survival

Diseased propagated materials help in the spread of the disease

Epidemiology

The disease is serious before flowering in the northwest region where the temperature is between 10-15°C during December-January. The disease is mild in the areas where temperature is between 15-20°C, sporadic between 20-25°C and nil beyond 25°C. The occurrence of malformation differed according to the age of the plants. 4-8 years old trees are highly susceptible.

Management

- Spraying with NAA at 100-200 ppm during October reduces the disease incidence.

- Eradication of malformed shoots and panicles after spring and autumn flushes (April and October)
- Spraying with acaracide (properguide 57EC) immediately after 3 flushing (February, May and October),
- Spraying with chelated copper (40 ppm) (mangiferin chelate or amino acid based chelate or copper fungicide) twice (August-September and December –January) before advent of the peak period of the fungal population,
- Spraying with chelated Zn++ twice (40 ppm) (December and February) to replenish the deficiency in the plants suffering long from the disease.
- *In vitro* test mangiferin Cu++ chelate killed the conidia and mycelia,
- *Aspergillus niger parasitized the Fusarium*
- While carbendazim arrested germ tube growth & reduced conidia production thus affected infection rate (r) of *F. moniliforme var. subglutinans.*
- Followed by spraying of Carbendazim 0.1% effectively controlled the disease.

2. Powdery mildew

Powdery mildew is one of the devastating foliar diseases of mango affecting almost all the cultivars. In India, the disease is wide spread including in the hill valleys and plains of U.P. and it is a serious threat to mango production. Its severity mainly depends on climatic conditions. The losses have been estimated upto 20% in Maharashtra and 30-90% in Lucknow and U.P.

Symptoms

A whitish powdery growth covers the stalks of the panicle, flowers and tender fruits. The whitish growth of the fungus comprising of asexual fruiting body an oidia. The affected flowers and fruits drop prematurely reducing the setting of fruits. At higher altitudes, the infection extends to the young leaves and twigs. Many of these are covered by the white powdery fungal growth and may exhibit distorted growth. On younger leaves it induces leaf curling.

Causal Organism: *Oidium mangiferae,* Berthet.

It produces septate mycelium and is hyaline, branched and superficial. Haustoria are sub epidermal. Conidia are hyaline, unicellular, elliptical and are borne singly or rarely in chain.

Mode of spread and survival

During off-season, the pathogen survived in intact green malformed panicles mostly hidden under dense foliage and its sexual forms has not been recorded.

During flowering period, the conducive environmental condition activates the dormant mycelium in necrotic leaves. Abundant conidia are produces and blown over to the flushes through the wind on young panicles which provides spore load for initiating the disease. Fresh infection on young leaves happens during first week of the December.

Epidemiology

It usually occurs during December-March. The disease is particularly destructive in the coastal areas during cold and wet seasons. Rains or heavy mist in the morning accompanied by cool nights during the flowering period favors the disease. Predominance of susceptible variety, high wind velocity for 3-4days with maximum temperature above 300c, minimum temperature around 150c and maximum relative humidity of 73.3-83.9% and minimum of 23.4-25.5% are found conducive for quick spread.

Management

The disease can be managed by pruning of diseased leaves and malformed panicles and three sprays of fungicides at different stages starting with wettable sulphur (0.2%) at the panicle size of 7.50 -10.00 cm followed by Dinocap (0.1%) after 15-20 days of first spray Wettable Sulphur (0.2%) can be used in all the three sprays and number of sprays may be reduced as per appearance time of disease.

3. Mango anthracnose

Anthracnose is also known as blossom blight, leaf spot, fruit rot and twig blight. This disease is severe both in field and storage. The disease is present all mango area of India The verities neelam and bangalora are highly susceptible to this disease.

Symptoms

Leaf spot

The fungus attacks tender shoots and foliage. Brown or dark circular or irregular spots are formed on the leaves and such leaves are crinkled. The affected portion dry up and fall off and leaf ragged margins. Often these leaves are shed leaving the twigs bare.

Month/Time of spraying	Stage of the crop	Fungicides	Quantity in 200 lit of water
October end or November start	For all the diseases before flowering	Copper oxychloride or Bordeaux mixture	800 grams or 1600g copper sulphate + 1600g lime
December	Emergence of flowering initial	Wettable sulphur 80% WP	600 grams
January	Complete emergence of inflorescence and formation of tiny fruits. For powdery mildew and anthracnose	Dinocap 48% EC or Hexaconazole EC5%	200ml 100ml 100ml
February	Pea size mango fruits. For powdery mildew etc.	Carbendazim 50% WP or Thiophanatemethyl 70% WP	100gram 100gram
March	Marble size mango fruits. For powdery mildew etc.	Dinocap 48% EC or Hexaconazole EC5%	200ml 100ml 100ml
April	Initiation of stone formation. For die back anthracnose	Mencozeb 75%	400gram 500gram

Die back

The infection spreads to the green twigs and forms dark brown lesions on them. Young branches dieback. On the lesions and dead portions, minute, pink, cushion-shaped fructifications of the pathogen are seen during moist weather.

Blossom blight

Small dark spots are formed on the main stalk and lateral branches of the panicle. Individual flower stalks are also infected. The flowers wither and shed. When severely infected. All the flowers destroyed and no fruits are formed.

Fruit rot

The tender fruits turn black and fall off. Often dark lesions develop on the fruits and cause partial of complete shriveling, blackening and shedding. Matured fruits are also infected. Black, round or irregular, sunken spots are formed on the skin as the fruit ripens the spots extend over the whole surface accompanied by the softening and rotting of the fruits. This type of injury is observed while the fruits are on the trees. It also occurs during transit and in storage. Spoilage of ripen fruits is common. Fructification of the pathogen is formed on the spots.

Causal Organism: *Colletotrichum gloeosporioides.* Penz, Spauld & Schrenk.

Acervuli developed on diseased parts of the plants. They are irregular and appear as brown to black dots. Setae are common on twigs but not on fruits. The acervuli when mature exude pink masses of conidia under moist conditions. Marginal setae are rare. Conidia are borne on hyaline conidiophores. The conidia are straight, cylindrical or oval, hyaline with two oil drops and are non-septate with round ends.

Mode of spread and survival

Inoculum remains on dried leaves, defoliated branches, mummified flowers and flower brackets and they serve as primary inoculum. Secondary spread is through air born conidia. The fungus can enter the pores of green fruits. The latent infection of mature fruits may takes place through lenticels. The fungus apparently infects the fruit while it is green and develops in flesh during ripening. The latent infection is carried from the field to storage. Healthy fruits develop infection after in coming in contact with disease ones. The latent infection does not begin to spread until it reaches eating maturity.

Epidemiology

The acervuli are abundant on the dead twigs and 80% of the spores on them are viable. Fresh acervuli continued to appear on dead twigs and persist on the tree. The optimum temperature for infection was found to be 25°C and relative humidity 95-97%. The perithecial stage of the fungus is not very common. There is no evidence to show that fungus perpetuates through ascospores.

Management

- Diseased twigs, leaves and fruits, which fall on the ground in the orchard, should be collected and all infected twigs should be pruned and burnt.
- Spraying of Bordeaux mixture 0.6% in the young plants during Feb, April and sept controls the disease. Spraying carbendazim 0.1% or thiophanate-methyl0.1% or chlorothalonil 0.2% for 15days interval until harvest effectively controls anthracnose.
- Before storage, fruits should be treated with hot water at 50-55c for 15min. Spraying of coc + zineb after completion of heavy showers followed by wettable sulphur 0.2% before flowering and carbendazin 0.1% at 15days interval from fruits formation proved effective.

4. Dieback

This disease is prevalent in all mango-growing states in India. In U.P.30-40% of road this disease affects side and other plantation.

Symptoms

The disease is characterized by dyeing back of twigs from tip downwards particularly in older leaves. It is giving an appearance of scorching by fire followed by complete defoliation. Barks are discolored and darkened at certain distances from tip. Such dark patches are generally seen on young green twigs. When the dark lesion increase in size, dying of young twigs begin at the base affecting leaf mid ribs extending along the veins. The upper leaves use their healthy green color and gradually turn brown accompanied by upward rolling of leaf margin. In, advanced stage, such leaves shriveled, fall off in a month are more, leaving the shriveled twigs. Internal browning in the wood tissue is observed on the slitting along the long axis. Cracks appear on branches, which exude gum. In fruits, the pericarp darkens near base of the pedicel. The affected area enlarges to form a circular, black patch, which under humid atmosphere extends rapidly and turns the whole fruit completely black within 2/3days. The pulp becomes brown and softer.

Causal Organism: *Botryodiploidia theobromae* Pat.

It is a Pycnidial fungi. Pycnidiospores are hyaline and thin walled, becoming thick walled, and dark brown and one septate. They have longitudinal striation and measures 20-30 x 10-15micrometer with paraphyses upto50micrometer long.

Mode of spread and survival

The fungus is a wound parasite. Dead twigs and bark of the trees harbor the fungus. The spores are spread through rain splashes.

Epidemiology

High temperature during summer predisposes the trees to the disease. Relative humidity of about 80%, max. &min. temperature of 31.5°C and 25.9°C c respectively and rains favour the disease development. It causes great damage and when mango grafts are kept in humid propagation shed.

Management

- Fruits should be harvested on clear dry days.
- Injuries should be avoided on fruits at all stages of handling.
- Care should be taken to prevent snapping off of the pedicel.
- Dipping of mangoes in 6% solution of borax at 43°C for 3min. gives effective control. Carbendazim 0.1% spraying in the field before harvesting gives effective control

5. Sooty mould

The disease is of common occurrence and affects many kinds of fruits and plantation crops.

Symptoms

The fungi produce mycelia, which is usually superficial and dark. They grow on the flowers, both tender and old leaves, stems and fruits. They grow and thrive on the sugary secretions of the plant hoppers and other insects. Black encrustations are formed on the surfaces of different parts of the plant. The photosynthetic ability of the plant is highly reduced because of sooty mould covering the photosynthetic area. During flowering time, its attack results in reduced fruit set and cause fruit fall. Black coating is also found on the fruits. Appearance of the affected fruits is lost and the price for such ugly fruits is usually low.

Mode of spread and survival: affected leaves and other crop debris serve as primary source of inoculum

Causal Organism: *Melioa mangiferae* Earle. *Capnodium ramosum* Cke. *C. mangiferae* Cke. & Brown *and Trichospermum acerinum (*Syd*).* Speg.

Epidemiology

High infestation with plant hoppers and the sugary substances (including excreta) secreted by them and other insects favour development of sooty mould. This is not a parasite or pathogen and do not draw any nutrient from the plants. Disease is severing in old and dense orchards where light intensity is low. Trees exposed to eastern side have fewer incidences while the trees in center of the orchard have more incidences. Continuous and heavy rainfall washes down these substances but high humidity proved congenial for growth of the fungus.

Management

- Both the insects and sooty moulds are to be simultaneously controlled in the eradication process. It is followed by spraying with a dilute solution of starch 1%. on drying, the starch comes off in flakes and the process removes the black mouldy growth fungi from different plant parts.
- Spraying insecticide followed by spraying with fungicide viz., Bordeaux mixture 1% is also recommended.
- Spraying of wettable sulpher, methyl parathion + gum acacia (0.2+0.1+3%) at 15days interval reduces the sooty mould incidence.

2. DISEASES OF BANANA

1. Panama Disease

Causal Organism: *Fusarium oxysporum* f.sp. *cubens* (E.F SMITH)

Panama wilt is one of the most devasting diseases of banana crop in the world being reported for the first time from Australia in 1874. In many countries banana trade was affected because of the wide spread occurrence of this disease, Rasthali, groups are susceptible. It is vascular pathogen.

Symptoms

Fungus blocks the vascular system and causes wilting. Yellowing of older leaves is the initial symptom. The infected plant shows yellowing of leaf blades developing as a band along the margin and spreading towards midrib. The leaf wilts and the petiole buckles. The leaf hangs between the pseudostems while the middle of lamina is still green. After four to six weeks, only the pseudostem remains, with dead leaves hanging round it. The cut stem smells like rotten fish. Pseudostem splitting is common in Fusarium affected plants. The plants from suckers growing out of diseased corms also wilt and the whole plants die.

Etiology

Mycelium-septate, sporodochium is asexual fruiting body. Asexual spores are micro and macro conidia borne on sporodochium. Sexual fruiting body is perithecium and it produces ascospores in the ascus. Vegetative, resting structures are chlamydospores.

Mode of spread and survival

Primary source of inoculum: The pathogen is both soil and water borne and also spreads through infected suckers (rhizomes) and survives in soil as chlamydospores for long period.

Primary source of inoculum: infected rhizomes. Secondary source of inoculum: Soil and water borne micro and macro conidia.

Epidemiology

Soil temp-28 -32°C, Relative humidity-85-90%. Acidic soil PH (5.5 to 6.0). Red loamy and sandy loam soil and Susceptible host.

Management

- Use healthy planting material. Collect the planting material from disease free area.

- Rhizomes are treated with dung solution and smear with Carbendazim powder.
- Use resistant verities like Robusta, grand naine.
- Based on soil pH apply lime@100-150gm per plant. Select nematode free soil, follow drip irrigation.
- Chemicals-carbendazim@1.0gm per lit. (as soil drenching) or capsule (carbendazim) insertion to the base of the rhizome or injection of carbendazim @ 10g/lit. to the rhizome.
- Bio control-*Pseudomonas fluorescens* 60mg per capsule. Each capsule applies to the corm. or application of *Trichoderma viride* to the soil along with FYM.

2. Moko Disease (Wilt)

Causal Organism: *Ralstonia solanacearum*

It was first recorded in Guyana in 1840. In India, the disease was first reported from West Bengal. Susceptible varieties are Robusta, Poovan from Tamilnadu

Symptoms

Younger leaves yellowing are the primary symptom. Gradually yellowing progresses downward leading to drooping and drying of leaves. Fruit bunch size gets reduced with immature and irregular ripening of fruits. Affected fruits shows cracking with bacterial ooze. Vascular browning of the fruit could be seen and light coloured vascular discolouration is common.

Etiology: Gram negative, lopotrichus bacteria. (More than one polar flagellum), multiplication by bacterial fission.

Mode of survival and spread

Primary source of inoculum: The bacterium is soil/water borne and also spreads through infected suckers/rhizomes. It survives in susceptible host like banana and heliconia.

Secondary source of inoculum: Bacterial cells spreads through irrigation water and also through suckers uses for planting

Epidemiology: Soil temp 28-32°C, Relative humidity 87-92%, pH slightly acidic to neutral, clay loam and sandy loam soil and susceptible variety.

Life Cycle

The bacteria survive through infected rhizomes and also in soil for 6 months to 2 year. The spread is through use of infected rhizome, cutting machetes at the time of planting, and through insects which carry the bacteria from oozing suckers and male flowers and bracts to healthy inflorescence and other parts of the plant. Entry into the host is mainly through wounds such as those caused during various cultural operations and during attack of insects and nematodes. The bacteria multiply rapidly in the xylem. Auxin balance of the plant is disturbed. IAA is synthesized by the bacterium and by the host and accumulates due to inhibition of the auxin degrading system. Loss of virulence in the bacterium is generally accompanied.

Management

- Use disease free planting material
- Use resistant verities like Robusta and Grand Naine.
- Affected plants should be collected and burnt.
- Give proper drainage and avoid movement of water from affected to healthy plants
- Drip irrigation method reduces the spread of bacteria.
- Chemicals: apply Copper Oxychloride @ 3gm per lit and Sreptocycline @ 0.5 gm per lit as a soil drench
- Use bio agent like *Pseudomonas fluorescens* @ 0.6%

3. Rhizome rot or Tip over

The disease occurs worldwide including India, Central America, Jamaica, and Israel etc. Generally the disease attacks the young plants and cause germination failure mostly in Cavendish and diploid group of bananas.

Symptoms: The Symptoms can be seen both externally and internally.

External symptoms: Failure of germination of planting materials when rotting of corm occurs extensively. Yellowing of leaves and stunting of plants. The trunk base may become swollen or split and results in the breakage of rhizome at soil level causing toppling of plants at the time of fruit maturing stage or during heavy wind periods. Sometimes the base of the trunk becomes dark in color with longitudinal splits. Diseased plants produce more number of suckers.

Internal Symptoms: Presence of pale grey water soaked area in the white cortical tissues at initial stage. This area then spread as finger like process inwards towards the centre of the corm, upwards up to the growing points and outwards up to the daughter suckers.The affected area in the corm become rottened, yellowish brown later on turns in to red color.

Causal Organism: *Erwinia chrysanthemi* and *E. carotovora* sp. Carotovora

Disease cycle

The pathogen enters into the rhizome through the wounds or decaying pseudostem leaf sheaths surrounding the central heart bud. The disease is spread mainly through suckers.

Management

1. Planting of healthy diseased free materials.
2. Improve drainage facilities, soil aeration, application of organic amendments and other cultural practices.
3. Growing of sunhemp as an intercrop for up to 5 months after planting.
4. Regulate irrigation.
5. Drenching the soil around the plant with either bleaching powder @ of 8g/lit or COC @ 1g/lit followed by drenching with carbendazim 0.1% and streptomycin sulphate 500 ppm conc. @ 2lit/plant two times at an interval of 10-15 days found effective in controlling the disease.

4. Sigatoka

First observed in Java, it has also been reported to cause severe losses in banana crop in countries like Columbia, Mexico, Jamaica, Panama, and India. In India Andra Pradesh, Assam, Bihar, Karnataka, Kerala, Maharashtra and Gujarat.etc. The disease was first detected in Java in 1902. The name "Sigatoka" was given when it was found in epidemic form in the *"Sigatoka valley" in Fiji island* in 1913. The disease is widely occurs throughout the banana growing countries. In India, it is prevalent in Assam, Bihar, West Bengal, Kerala, Tamil Nadu, Karnataka, Maharashtra and Gujarat.

Causal Organism: *Cercospora musae (Zimm.)*

Symptoms

Light yellow or brownish green narrow streaks which enlarge in size developing into linear, oblong muddy brown to black spots. In central portion necrotic circular spots on older leaves, this spot leads blight and splitting of leaf lamina. Then complete leaf drying followed by defoliation Yellow streaks and spot later turn brown and enlarge to form yellow spots with greyish white center.

Etiology

Conidia are elongated, narrow and multi septate borne on conidiophores. Perithecia is dark brown to black. Asci are oblong. Ascospores is septate. Hyaline, obtuse ellipsoid with upper cell slightly broader. Ascospores are sexual spores.

Mode of survival and mode of spread: Primary source of inoculum: dormant mycelia present in affected debris.

Secondary source of inoculum: Air borne conidia

Epidemiology: Warm temp 23 - 25°C, rainy or humid weather. Poor or badly drained soils and in shady areas. Closer spacing, heavy Weed or grass cover and neglected crops.

Life Cycle

The pathogen can survive on dry infected leaves on the field soil. It is spread through conidia and ascospores as air borne. Conidia are formed in humid weather throughout the year but their release and germination depends on water or high humidity. They are dispersed by rain drop splashes and by wind. Ascospores are shot out violently through the ostiole in response to wetting of perithecia and are dispersed by air currents. They are responsible for long distance spread of the pathogen while conidia are generally the most important means of local spread. The infection by both types of spores produces the same type of spots and subsequent development of the disease. Sigatoka spreads fast in a humid weather or high rainfall. Little infection occurs at temperatures 21°C- even if humidity is optimum. In dry weather with high day temperature and little dew during night the disease fails to spread. Soils with poor drainage and low fertility favor the disease. Conditions which are conducive for increased humidity in the plantation are favourable for the disease. Thus thick planting, presence of weeds and increased number of suckers promote disease development.

Management

- Removal and destruction of affected leaves followed by spraying with Bordeaux mixture 1% +linseed oil 2% is recommended
- Spray with oil based copper fungicides is also found effective. Carbendazim @ 0.1 % or Mancozeb @ 0.25 % spray with spreading agents like teepol is also recommended
- Propiconazole @ 0.1% spray with spreading agent effectively manage the disease

5. Anthracnose/ Fruit Rot

It is serious disease in banana growing areas especially in Bihar, Karnataka, Tamil Nadu. Almost all varieties are susceptible, but severity may vary, owing to severe infection on table varieties. The disease is also called by different names viz., stem end rot, neck rot, back end, finger stalk rot etc.

Causal Organism: *Colletotrichum musae* (Berk and Curt)

Symptoms

It can be seen in the distal end of banana. The skin turns black. Shrivels and covered with characteristic pink colored asexual fruiting body acervulus. As the disease advances, it spreads to entire finger, entire bunch and resulting in premature fruit ripening. The shrivelled fruits covered with pink spore masses, which finally rot. Ripe fruits are more susceptible than unripe fruits. **Latent Infection:** usually originated in the field on uninjured fruits. When fruits approaches maturity, the fungus resumes activity and cause typical lesions on ripe fruits.

Non-latent Infection: usually begins during or after harvest as small peel wounds and continue to develop without dormancy.

Etiology

Asexual fruiting body is Acervulus. Conidiophores are cylindrical, septate, branched and sub hyaline towards the base. Conidia are hyaline, aseptate, oval to elliptical, flattened at the base.

Mode of spread and survival

Primary source of inoculum: The fungus survives as dormant mycelia for long time in the fallen leaves. Secondary source of inoculum: Air and splash borne conidia produced from the acervuli.

Epidemiology

The disease is favored by high temperature 30 -35°C and relative humidity 85-100% and also by fruit damage. Black end is the name given to the decaying of stem end on single fingers whereas, Finger stalk rot are also known as Santa Marta stem end rot or Neck rot and is common in complete bunches. Disease is more abundant during rainy season. Ripe fruits in storage are more susceptible than the unripe fruits in the fields. Cavendish is more susceptible variety.

Management

- In the field, distal bud should be removed when all the hands have opened to prevent infection.
- Infected materials must be burnt.
- Fruit should be free from infection and as found as possible before it is transported, stored and ripened.
- Banana bunches should harvested at correct stage of maturity. After harvested of the bunches, they should be transported to the storehouse

without causing any bruises to them. The transported bunches should be carefully and cooling should be done.

- Fruit stored at 7-10°C.
- Pre harvest spray with Carbendazim@ 0.1% four times at fortnightly interval is highly effective.

6. Bunchy Top

Also called as cabbage top disease, first recorded in Fiji in 1889. Cavendish banana and local plantations. Around 1940, bunchy top disease was observed in Indian states Assam, Bombay, Kerala, and Tamilnadu.

Causal Organism: *Banana bunchy top virus*

Symptoms

The affected leaves shows green streaks on the secondary veins on the under side of lamina and on the midrib and petiole. The powdery bloom covers the midrib and petiole, if this is rubbed off the dark green line with a ragged edge. Leaves become dwarf, they also show marginal chlorosis and curling. The leaves are brittle in texture petioles are incompletely elongated. The leaves become smaller and eventually the crown of the plant become composed stunted growth, the rosette or bunchy leaves. Fruit bunches are reduced, the lower hands of the bunch often die off.

Etiology: Semi persistent in nature, aphids are act as vector (*Pentalonia nigronervosa* f.sp. *cocquerelli).* Virus particles are isometric and are 18-20 nm in diameter. Acquisition feeding period 2- 4 hours, incubation period 6-8 hours, inoculation period 2-4 hours.

Mode of spread and survival

Primary source of inoculum: Affected plants and affected suckers.

Secondary source of inoculum: Vector borne virus particle, spread more in summer through aphids.

Epidemiology

The disease is spreading very quickly in hill zone, because the hill banana is cultivated as a perennial crop and the aphid population in the hill areas is more due to conducive atmosphere in these areas.

Management

- Selection of healthy planting materials

- Regular field visits and removal of affected plants and suckers at periodical intervals
- Select disease free area for new plantations
- Avoid excess application of nitrogen, apply recommended dose of potassium, FYM.
- Proper drip irrigation in summer season is good
- For aphids management, application of imidacloprid 17.8SL @ 3 ml/10lit. water through inject banana by injector.
- Spray Afidopyropen 5DC @20ml/10 lt. of water.

3. DISEASES OF GRAPES

1. Powdery Mildew

Powdery mildew is an endemic disease wherever the grapes are grown in the world. The disease has been reported from the American continent, Europe, Africa, Australia and Asia. In India, the disease is most common in Maharashtra, Gujarat, Andhra Pradesh, Karnataka and Tamil Nadu.

Economic Importance

The disease causes extensive damage in whole of Europe and Western USA, sometimes destroying the crop completely. French grapevine industry also suffered huge losses due to the epidemics of this disease during 1850-55.The disease not only reduces the yield and lower the fruit quality but wine prepared from infected fruits often develops off-flavor (Ough and Berg, 1979).

Symptoms

Fungus attacks all the green parts of the plant at all stages of plant growth. The fungus produces white to grayish powdery patches on the affected plant parts including fruits but young leaves are most susceptible and develop small whitish patches both on upper as well as lower surface. These patches grow in size and coalesce to cover large areas on the leaf lamina and gets twisted. Malformation and discoloration of the affected leaves are also common symptom, resulting in distortion. Similarly, powdery patches are also produced on the stem, tendril, flowers and young fruit branches. Diseased vines appear wilted and the stem portion turns brown. The infected blossom and berries turn dark in colour, irregular in shape and brittle. In advance stage of infection, berries may develop cracks and such berries do not develop and ripe. When blossom is affected, flowers may drop off. Affected berries become malformed and skin cracks and pulp may be exposed under such conditions.

Causal organism : *Uncinula necator* (Schw.) Burr., (Syn. *Oidium tuckeri* Berk.) which is an obligate parasite.

Etiology

Mycelia septate, external thin mycelia, haustoria sub epidermal, obligate parasite. Asexual spores are: barrel shaped, conidia borne on oidiophore in chains. Asexual stage of the pathogen is oidium. Sexual spores: Ascospores inside the Ascus in the ascocarp which is called as Cleistothecium.

Mode of Spread and Survival

It survives as dormant mycelia and as Cleistothecia on the shoots and buds from season to season. The disease spreads by the air- borne conidia/oidia.

Epidemiology

The disease occurs in severe form from Oct- Nov in North India and Feb- June in South India. Disease is favoured by warm sultry weather and retarded by sunshine. Warm winter temperature from 20 to 33.50C has been found to be the cause for epidemic in Hyderabad. Disease development is adversely affected by rain.

Management

Cultural practices

- The use of training systems which allow proper air circulation through the canopy and prevent excess shading helps in reducing the disease.
- Orchard sanitation is also important in reducing the disease pressure during the growing season.

Chemical control

- Fungicides like Sulphur, Dinocap, Triadimefon, Penconazole, Mycobutanil and Flusilazole are used commercially although not as extensively as sulphur, to control the disease.
- The use of fungicides for control of powdery mildew should begin during early stages of vine development. Spray schedules at an interval of 7-10 days are usually required for effective control by sulphur.
- Dinocap is to be given at an interval of 10-14 days while; sterol biosynthesis inhibiting fungicides are commonly used at 14-21 days schedule.
- For effective control, the fungicide spray should start just after bud burst.

- The fungicides should be sprayed alternatively and the same fungicide should not be sprayed continuously.

2. Downy mildew

Downy mildew is most destructive fungal disease of grapevine and occurs in most grape growing regions of the world. Before 1870 the disease was endemic to USA. It was first reported in Europe in 1878, within short period of time, the disease spread like wild fire in France and posed a threat to the vine industry. It was during this time that Prof. P.M.A. Millardet discovered the Bordeaux Mixture against downy mildew. He observed that the vines near to road side, sprayed with lime and copper sulphate mixture to avoid pilferage by street goers, were quite healthy as compared to the vines in the interior of vineyards (Millardet, 1885). Introduction of downy mildew to various countries was probably by way of infected nursery/propagating stocks.

Economic Importance

The disease results in cluster destruction and loss of vine foliage or photosynthetic area. Almost entire crop fails whenever the conditions are conducive for disease development. Under epidemic situations, vines may be defoliated, which results in nakedness of canes and exposes fruit to sunburn. In the next season, vine vigour and crop potential may be reduced.

Symptoms

Symptoms of the disease appear on all aerial parts of the plant. The disease is usually first observed as small translucent, pale yellow spots with indefinite borders on the upper surface of leaves. Whitish downy growth on the lower surface of the leaves comprises of tufts of mycelia, sporangiphores and sporangia of the fungus. On the corresponding upper surface, small round to angular light green/chlorotic patches is the characteristic Symptoms. On the under surface of leaves and directly under the spots, a downy growth of the fungus appears. The tissue in the spot is traversed by reddish lines. Later, the infected areas are killed and turn brown .The growth on the lower surface becomes dirty grey. Tender vines are also affected. Infected leaves turn yellow, brown and wither. Flowers die and drop off. Fruits become grayish, skin become hardened and shrivels resulting in mummified berries. The necrotic lesions are irregular in outline and they enlarge and coalesce to form larger necrotic areas on the leaves, frequently resulting in defoliation. Diseased shoots remain stunted. Infected leaves, shoots and tendrils are covered with whitish growth of the fungus. Flowers and berries are also affected. Flowers may blight or rot. During blossom or early fruiting stages, entire clusters or part of them may be attacked and become quickly covered with the downy growth and die. If infection takes place after the berries are half-grown, the fungus grows mostly internal. The berries become leathery and wrinkle and develop a reddish marbling to brown coloration. The fruit shed if the attack is very severe. The juice quality of fruit is found to be reduced. Infection of green young shoots, tendrils, stems and fruit stalks results in stunting, distortion and thickening of the tissues. Infected tissues turn brown and die.

Causal Organism: *Plasmopara viticola* (Berk. & Curt.) *de Bary*

Etiology

Mycelium is intercellular with spherical haustoria, coenocytic, thin walled and hyaline. Sporangiophores arise from hyphae in the sub-stomatal spaces and sometimes they emerge directly through the cuticle. On young berries of grapevine they emerge through the lenticels. Sporangiophore branching is almost at right angles to the main axis and at regular intervals. Secondary branches arise from lower branches. From the apex of each branch 2-3 sterigmata arise and bear sporangia singly. The sporangia are thin walled, oval or lemon-shaped. Asexual spores: Zoospores are pear shaped, biflagellate borne in sporangia. Sexual spores: Oospores thick walled, diploid developed through gametangial contact (union of dissimilar gamets, Oogonium and Antheridium) and are also called as dormant spores.

Mode of spread and survival

The pathogen survives on the infected leaves and vines as oospores and dormant mycelium. The secondary spread is through wind- borne sporangia and zoospores which are found on the new flush.

Epidemiology

The most favourable temperature for germination of sporangia is between 10-230C. Disease development is favoured during rainy season when there is heavy dew, relative humidity is above 80% and temperature is between 23 and 27°C

Management

Regulatory measures

- Restriction on the movement of planting material at regional, national or international level should be imposed, since the pathogen spreads through dormant cuttings (planting materials).

Cultural practices

- All infected plant material and pruned parts must be removed and burnt before bud sprouting, so as to reduce primary inoculum.
- Even during growing season plant debris must be avoided in and around the field.
- Careful attention should be paid to spacing of vines, row direction and placement of wind breaks, which will ensure maximum air drainage and minimum leaf wetness duration.
- To encourage air movement within the plant canopy, practices such as, removal of leaves around berry clusters, trellis designs and pruning systems which allow more air movement be followed during the early development of vines.
- Careful disbudding and training of veins should be practiced to maximize distance between soil and foliage.

Chemical control

- After pruning, the vines should be sprayed with Bordeaux Mixture 1.0% or Difolatan 0.2% or Copper oxychloride 0.3% or Chlorothalonil 0.2%.
- When the flushes are formed, spraying with Difolatan 0.2% or Chlorothalonil 0.2% or Metalaxyl 0.2% or Copper oxychloride (%) are effective.
- It may be repeated at weekly or fortnightly intervals depending up on severity and weather conditions.
- When the non-systemic fungicide is used during humid and rainy period spraying should be repeated for every two or three days.

Biological control

- *Erwinia herbicola,* a saprophytic bacterium, used as liquid culture and sprayed on vines which inhibits *P. viticola* upto 75%.

3. Anthracnose/birds eye spot

Anthracnose is a widespread disease in all grape growing regions of the world. The disease is known in Europe since ancient times, however, in India, the disease was first recorded in 1903 near Poona and is now widely prevalent in Rajasthan, Uttar Pradesh, Punjab, Haryana, Andhra Pradesh, Karnataka and Tamil Nadu. Under north Indian conditions, the disease appears only during rainy season. In the southern part of the country, the berries escape infection because the crop matures before the onset of rains

Symptoms

The fungus attacks shoots, tendrils, petioles, leaves, veins and stems and also inflorescence and berries. Numerous spots occur on the young shoots. These spots may unite to girdle the stem and cause death of the tips and may also cause die-back Symptoms. Spots appear on the new shoots and fruits also. Spots on petioles and leaves cause them to curl or become distorted. On berries, characteristic round, brown sunken spots resembling “Birds Eye” and hence the name of the disease. On the leaves it appears as small, irregular, dark brown spots. The central tissue turn grey and falls off. The disease appears as dark red spots on the berry. Later these spots are circular, sunken and ashy grey and in late stages these spots are surrounded by a dark margin which gives it the bird s eye appearance. The spots are 7mm in dia but they may involve about half of the fruit

Causal Organism: *Gloeosporium ampelophagum (Pars.) Sacc. {Elsinoe amphelina* (de bary) Shear –perfect stage*}*. Conidial stage is *Sphaceloma ampelina* de Bary . Conidia are formed in pink acervuli. They are hyaline, single celled, oblong or ovoid. Perithecia (Pseudothecia) are small and unconspicuous. Asci are globular and ascospores are hyaline, 3-celled.

Mode of spread and survival

The pathogen survives as dormant mycelium in the cankers on the stem and on the infected twigs. Secondary spread is through conidia which are carried by wind and rain water.

Epidemiology

The disease is severe during July- Aug and Nov- Dec months. Infection in new sprouts takes place during rainy season. Heavy rains after pruning leads to more incidence.

Management

Cultural practices

- Training of vines should be such that water splashes should not reach the foliage, canes and branches during rainy season
- Ground level canes and branches should be removed. All cankerous canes should be pruned and destroyed by burning.
- This will help in reducing the primary inoculum during the growing season

Chemical control

- Spray vineyards at the time of leaf emergence with Thiophanate methyl (0.1%), bitertanol (0.1%), carbendazim (0.1%), or Bordeaux mixture (1.0%).
- At least four sprays of fungicides should be given during rainy season at fortnightly intervals. Care should be taken not to spray the same fungicide regularly.

Varietal Resistance

- The Muscadine grapes seem to be immune, Champane highly resistant, Concord moderately resistant and most varieties are highly susceptible.
- Variety Delight is tolerant whereas, Bharat Early and Hussaini are resistant.

4. Bunch rot

Bunch rot, also known as grey mould rot or *Botrytis* rot and is prevalent throughout the world wherever the grapes are grown. The maximum damage of this disease is noticed in berries at the harvest time as well as during transport and storage.

Economic Importance

Infection on flowers and berries is most important from economic point of view as it lowers both quality and quantity of fruit. The flowers provide an excellent source of nutrition to the fungus. The berries are resistant to infection during development stage until maturity after which, these become increasingly susceptible. The famous noble rot represents a rare case of rotted food stuff becoming more valuable than healthy one. Under favourable dry conditions, following are heavy *Botrytis* attack, the mycelium colonizes the berry skin and kills epidermal cells thereby allowing abundant evaporation of water through cuticle. After drying, a raisin-like shrunken fruit is picked up selectively. The famous white Auslese- type wines, the most renowned originating from the Rhine Valley or Sauterne are made from such grapes.

Symptoms

The disease Symptoms appear on all plant parts i.e. leaves, shoots, flowers and berries. Both young and relatively older leaves are infected by the pathogen. The fungus produces irregular, necrotic spots in the centre of the leaves. Under certain conditions, marginal necrosis also occurs. Infected flowers normally do not develop any apparent Symptoms but necrosis of stamens, the solitary ovary can often be seen covered with tufts of sporulating mycelia. The most prominent Symptoms of the disease are found on the berries. Infected berries become dark coloured and show typical grayish, hairy mycelium all over their surface. Tufts of conidiophores and conidia protrude from stomata and peristomatal cracks on the skin of the berry. Under high disease pressure, all the berries in a bunch gets affected.

Causal Organism: *Botrytis cinerea* Pers. ex. Fr.

The fungus produces grey growth on the surface of the fruit but in high humidity, the mycelial growth may be cottony and white. The conidiophores are long, slender, erect, hyaline, unbranched or seldom branched. The epical cells enlarged or rounded bearing clusters of conidia on short sterigmata. Conidia are hyaline, ovate or elliptical to almost globose, one celled conidia appear grey in mass. Black irregular sclerotia are frequently produced.

Mode of spread and survival

The fungus, *Botrytis cinerea* survives from season to season on the grapevines, rotted berries and stem clusters in the form of mycelium, conidia and sclerotia. The conidia of *B. cinerea* are dry and largely dispersed in air currents.

Epidemiology

The optimum temperature for sclerotial germination followed by infection is between 20 and 25°C and in relatively dry soil. Sclerotia are more likely to survive longer on canes and these are probably more important than those in soil as are source of primary inoculum.

Management

- Maintaining the sanitation in the vineyard is the most important cultural practice to keep the disease under check.
- Diseased vines, leaves and fruits must be picked up and destroyed.
- Removal of grape mummies acting as primary sources of infection from vines at the time of pruning and burning them.
- Fungicides like dicarboximide, procymidone, vinclozolin and iprodione are effective in disease control.
- Removal of leaves in the vicinity of flower clusters and bunches helps in reducing the disease severity.
- Some of promising new botrycides in grapes are triazole, folicur, sterol biosynthesis inhibitor (SBI).

5. Black rot

The disease after its introduction in France during 1880 s, it spread to all grape growing areas of Europe. In India, the disease has been observed in Madurai district. Black rot is more distructive in warm and humid areas than in the cooler and drier ones. The disease has been recently observed on certain purple varieties and it is less common on the seedless and Pachha draksha varieties. The disease on fruit begins to show as light, brownish, soft, circular spots which increase in size and the entire berry is discoloured. The decaying berries begin to shrivel within a week and are transformed into hard, black, shriveled mummies. On the leaf, circular red spots appear and later the margins become black. Minute black dots representing fruiting bodies of the fungus are arranged in a ring near the outer edge.

Causal Organism : *Guignardia bidwelli (Ell.)* Viala & Ravaz.

The mycelium is hyaline when young and it becomes brown after full maturity. Perithecia are globose, ostiole not prominent. Asci are clavate, thick walled. Each Ascus contains 8 ascospores. Ascospores are bicelled but cells are unequal in size. Ascospores are hyaline, sub-ovoid or elliptical, slightly flattened on one side.

Mode of spread and survival

Perithecia develop on mummified grape berries and the Ascospores are discharged when mummies are wet. Ascospores produce germ tube and penetrate directly through the cuticle. Primary infection occurs on young leaves and fruit pedicles. Pycnidia are rapidly produced. Pycnidiospores spread through meteoric water. They may survive the winter and germinate in the following season.

Epidemiology

Frequent rains and humid climate are conducive for disease development.

Management

- Diseased berries and leaves should be collected and destroyed.
- Spraying of Bordeaux Mixture 1.0% or Ferbam 0.2% or Captan 0.2%, Chlorothalonil 0.2% should be done when the new shoots are 15-25cm long and repeated before bloom, 10-15 days after bloom.

4. DISEASES OF CITRUS

1. Citrus gummosis/ Leaf fall/ Foot rot: *Phytophthora citrophthora*

Symptoms

This is soil borne fungi. Primary colonization is on roots as discolouration, root decay, bark regradation at collar region and leaf falling. The exudation of gum like substance from the bark of the trunk, the bark cracks open and in the later stages dries up and fall off.

Etiology

Aseptate mycelia, zoospore are asexual spores produced in the sporangium, oospore are sexual spores or resting/dormant spores borne in oogonium

PSI: Dormant mycelia and oospore present in effected debris and infested soil.

SSI: Zoospore spread through soil, irrigation water.

Epidemiology

Cool weather, temperature 18-22°C: 90-95% RH, High soil moisture, PH 6-7.

Life Cycle

There are 3 stages of life cycle:

1. Asexual Stage: Zoospores borne in sporangium
2. Sexual Stage: Oospores borne in Oogonium and
3. Vegetative Stage: Mycelia with haustoria

Oospores are sexual spores and also resting spores, present on affected debris for a longer time (6-8 months). When the conditions are favourable, these oospores germinate by producing germ tube, the tip of the germ tube swells to form sporangium. Initially sporangium is multinucleated structure, then each nuclei starts formation of zoospore wall. Once these zoospores matures, they start moving randomly and burst open the sporangium wall and become air borne. Air borne zoospores move certain distance, then they loose their flagellum and forms circular which is the encystment of sporongia. Haustoria is intercellular, Mycelia is intracellular. Once the conditions are adverse temperature increases, dried humidity, due to this fungi switch on to the sexual reproduction, here male reproductive organ is Antheridium and female reproduction oogonium between 2 gametangial processes. Oogonium is circular in nature, eunucleated, sometimes 1 cell or 4 cells are there. Antheredium tubular in nature and multinucleated, Gametangial contact.Once union of oogonium and gametangial takes place Plasmogamy takesplace. After Antheridium lesicata takes place karyogamy.

Management

- Provide good drainage, thereby it creates adverse condition and asexual reproduction reduces and inoculum decreases
- Uproot severly infected plants, Replant with tolerant varieties.
- Application of *Trichoderma* (100 g per plant.)
- Chemical soil drenching of Bordeaux mixture 1%, Copper oxy chloride 3gm per lit. of water, and aerial spray.
- Use resistant root stock for grafting.
- Avoid low lying areas for citrus.
- Avoid exess N application,apply recommended K Application

2. Powdery mildew: *Oidium tingitaninum*

Symptoms

Whitish powdery growth on young leaves & twigs. The affected leaves get distorted and in severe condition drop down. Infected twigs exhibit characteristic die back symptom. Young fruits are also covered by whitish powdery mass of the fungus and drop off prematurely, resulting in poor yield.

Etiology and Spread

Comparatively cool and moist regions are prone to disease development. Damp mornings with are few hours of sunshine favour onset of the disease. The fungus is an ectoparasite and absorbs food materials from the epidermal cells of leaf through houstoria. It is a wind- borne disease. Septate mycelia, barrel shaped conidia born in chains, ectophytiic, sub epiderml haustoria, external mycelia. PSI: Dormant mycelia. SSI: Air borne barrel shaped conidia.

Life cycle

Dormant mycelia present in the affected parts. During congenial conditions germinates and produce oidea. After maturity barrel shaped conidia releases, flight and land on host. Infection takes place by producing sub epidermal haustoria & plant start producing powdery growth comprising of oidea. Oidea is an asexual fruiting body of the powdery mildew, barrel shaped conidia borne in chains on oidiophore. Then they release, flight & landing on their respective host. Infection process continues asexually.

Management

- Prophylactic measures: cloudy warm weather, spray wettable sulphur 3gm/ lit. of water

- Aerial spray: carbendazim 1.25gm/ lit. of water
- Wider spacing
- Avoid high density planting
- Avoid excess N application
- Provide recommended K application

3. Citrus scab/ verrucosis: *Elsinoe fawcettii* and *Sphaceloma fawcettii*

1. Commons cab or sour orange scab-*Elsinoe fawcettii*
2. Sweet orange scab-*E. australis*
3. Tryone scab-*Sphaceloma fawcettii var. scabiosa*

Symptoms

Whitish, raised, circular, scabrous growth on the fruits, later the color turns to grayish color, decreases the fruit size, quality and fruit fall off. Leaves: on lower surface of leaves whitish scaborous growth corresponding upper surface, concave dippression can be seen.

Etiology and spread

It is believed that the pathogen perpetuates and survives in off season as perithicium. Secondary spread may be through the conidial stage, which is mostly produced on the host. Conidia are produced between 7 °C & 33 °C at 66-100% RH on young lesions. Conidia from old lesions are dispersed during rains, but only to short distance.

Management

1. Collect the infected leaves and burn it.
2. The disease can be controlled by spraying with 1% Bordeaux mixture and difolatan.
3. Chemical:Carbendazin-1.25gm/lit
4. Avoid excess N application
5. Provide recommended K application

4. Citrus sooty mould: *Capnodium citris*

It is not actually a disease of plants. The fungi purely grows on the surface by utilizing the insect excreta or honey secretions by insects and plant. By growing such blacky mold on the surface, abstructing the sunlight to reach the photosynthetic area (green chlorophyll) of the plant and thus interfering in photosynthesis.

Symptoms

Black colored sooty mass covering the leaf surface, sometimes on young stem, fruit surfaces. Black sooty mass comprising of conidia and mycelia. Affects normal photosynthesis, thereby plant growth decreases. This is purely ectophytic and not plant parasitic fungi. By utilizing leaf exudates and honey like substances secreted by insects and also insect excreta, this fungi grows on the surface.

Management

1. 1% Starch sprays, after it forms flakes on the sooty mass, along with flakes sooty mass fall off from the leaves after drying.
2. Spraying systemic insecticides to manage the insects population could help in avoiding or reducing the sooty mold.

5. Citrus anthracnose: *Colletotrichum gloeosporioides*

Symptoms

The disease leads to defoliation and tip drying of twigs, it is called whither tip. Shedding of leaves and dieback of twigs. On the dead twigs acervuli appear as black dots. Light green spots appear which later turns brown. The pathogen also infects the stem-end of immature fruits causing fruit drop. In severe cases branches die back.

Etiology: Septate mycelia, asexual fruiting body-acervulus setae are present.

Primary source of inoculum: Dormant mycelia

Secondary source of inoculum: Conidia produce by Acervulus

Epidemiology: Warm weather, temperature 30-32°C, RH 80-85%, Cloudy weather susceptible to host.

Management

- Collect the affected leaves and burn it.
- Avoid excess N application.
- Summer irrigation is best.
- Chemicals: Carbendazim-1.5g/L.

6. Citrus canker/ Bark erruption: *Xanthomonas campestris* pv.*citri*

Symptoms

Leaves: Initially water soaked patches, these slowly turns to brown discoloration later produce corky raised spots then leads to yellow hallow.

Stem: Same as leaves but no yellow hallow, bark eruption takes place, from cracks we can see bacteria ooze during warm rainy season.

Fruits: brownish corky out growth and cracks formation and later crater like appearance is the common symptom. Marketing quality reduces, fruit size reduces.

During preservation it leads to rotting. Older lesions will turn tan to brown and have a yellow halo surrounding the raised margin on leaves. Lesions on stems and fruit appear dark brown to black. Leaves infected by citrus leaf miner are more susceptible to infection by the canker bacterium.

Etiology and spread

Canker-infected leaves, twigs serve as the source of inoculum to spread the disease from season to season. However, the cankered leaves drop off early and bacteria perish rapidly in the soil.

Primary source of inoculum: Affected plant, soil

Secondary source of inoculum: Bacterial cells spread through Irrigation water, Agricultural operations, pruning shears.

Epidemiology: Prevalence of 20-35°C temperature, high humidity and the presence of moisture on the host surface.

Life cycle

The bacterium enters the host through stomata or wounds. It multiplies in the intercellular space, dissolves the middle lamella and establishes in the cortex region. Canker pustules develop an exude bacteria in the form of gummy substance. They are freely disseminated, chiefly by wind and Considerable extent by rains. Citrus leaf-miner helps dissemination and infection of citrus canker. Leaves affected by miner and canker get distorted and drop off early. The injury to the leaf epidermis made by the borrowings of leaf-miner serve as an easy opening to the canker bacterium and the canker lesion appear throughout the zig zag manner.

Management

- Quarantine: If area is disease free, restrict the entry of planting material from infected to healthy area.

- Cultural: affected leaves, stem, fruit cut and burn, cut end portion of stem, paste with Bordeaux paste.
- Hot water treatment root stocks 50°C for 10-15 min.
- Biological: *Pseudomonas fluorescens*
- Chemicals: Spray 1% Bordeaux mixture
- Spray 0.3% Copper Oxychloride + 500 ppm Streptocyclin in 1 lit of water

7. Citrus Tristeza/Quick decline

Symptoms

Leaf: Chlorosis is the common symptom, leaf size reduction, leaves drop off and defoliated twigs die back.

Stem: Bark eruption and pittings on stem (v shaped dipression on the stem and stem twisting occurs.)

Fruits: In the affected fruits thicknesss of the rind is increased, mesocarp decreases.

Root stocks are susceptible, phloem necrosis is the common symptom, root discoloration and root decay takes place it leads to sudden leaf fall.

Spread

Primary Source of Inoculum: Affected plants, affected cuscuta

Secondary Source of Inoculum: Vectors (aphids) (*Toxoptera citricida*), mechanically sap/ grafting/budding

Management

- Use the seedlings obtained from healthy seeds for transplanting.
- Use a rough lemon root stock and other scion, protecting plants from phloem necrosis.
- As the disease severity increases cut and burn the affected parts.
- Phased manner replanting with resistant plants.
- Hot water treatment of rootstocks at 45 °C for 10-25 min.
- Removal of cuscuta and spraying of systemic insecticide, Dimethoate 2ml/lt or Acephate 75% SP@1 gm/Litre of water or Deltamethrin 2.8% EC@1 ml/Litre of water controls the aphid vector population.
- Apply recommended dose of N P K and FYM.

- Heat treatment.
- Cross Protection: Use the pre-immunized seedlings with mild strain of the virus to manage the disease and to avoid the losses.

5. DISEASES OF GUAVA

1. Fusarium wilt : *Fusarium oxysporum* f.sp. *psidii*

Occurrence of serious wilt was reported from Haryana, Punjab, Rajasthan, Uttar Pradesh and West Bengal.

Symptoms

The disease is characterized by yellowing and browning of leaves, discolouration of the stem and death of the branches along one side. Sometimes the infection girdles the stem and the whole plant may wilt. Leaves die and the twig barks split.

Pathogen: *Fusarium oxysporum f.sp. psidii* Prasad, Mehta and Lall. F.solani (Mart.) Sacc., Mycelium is white or pink with a purple tinge. Microconidia are borne on simple phialides arising laterally on the hyphae. Microconidia are oval to ellipsoid, cylindrical, straight to curved and 7 to 10 x 2 to 3 pm. Macroconidia are 3 to 4 septate and 32 to 50 x 3 to 7 um in size. They are fusoid to subulate and pointed at both ends. Sporodochia and spinanodes are present. Chlamydospores may be intercalary or terminal.

- Asexual spores : Micro & macro conidia
- Vegetative spores : Chlamydospores (Resting spores)
- Sexual spores : Ascospores borne in ascus

Mode of spread and survival: The fungus first colonizes on the surface of the roots and enters the stem tissues at the basal portions near the ground level. It multiplies in vascular region and affects the cortical cells.

Primary source of inoculum: Soil borne inoculum in the form of chlamydospores and infected plant parts.

Secondary source of inoculum: inoculums produced on the infected host

Epidemiology: Higher disease incidence is noticed during the monsoon period. The disease appears in August and increases sharply during September - October. It is severe in alkaline soils.

Management

- Dry branches should be cut off and wilted plants uprooted.

- Soil should be treated with lime or gypsum to make the soil pH 6.0 to 6.5 balanced nutrition of host reduces seventy of the disease when organic nitrogen is supplied.
- The soil of the pits should be treated with 37 to 40 per cent formaldehyde (45ml of formaldehyde plus 270 ml of water plus 35kg of soil).
- This treatment has to be covered with a polythene sheet for at least 15 to 20 days. When the traces of formalin disappears, the pits are filled with this soil after planting the tree.
- Soil drenching of Carbendazim 1.5g/lit considerably reduces the disease.

2. Fruit canker/Scab/Grey blight: *Pestalotiopsis psidii*

Symptoms

Infection generally occurs on green fruits. Minute, brown or rust- coloured, unbroken, circular, scabby lesions of 2 to 4 mm dia appear on the fruit which later tear the epidermis open in a circinate manner. The margin of the affected area becomes raised. The scab disfigures the fruits and their market value is highly reduced. Primary source of inoculum: Dormant mycelia. *Helopeltis antonii,* a kajji bug which punctures the young fruit and suck the juice and that damage exposes the fruit to infection by the pathogen.

Mode of spread: spread is through the wind-borne conidia.

Epidemiology: The fungus is capable of growing at temperature between 20 and 25°C. Mycelial growth with intensive sporulation takes place at 5.5°C. Wounding results in quick attack by the fungus.

Management

- Since the wound by insect predisposes the fruit to infection, spray the young fruits after pollination with a suitable systemic insecticide (Dimethoate – 2ml/l) will take care of the infection.
- Spread of the disease can be checked by three or four spraying with Bordeaux mixture1.0 percent or copper oxy chloride 0.2 per center.
- Summer irrigation +Nutritional Management reduces the disease.

3. Stem canker: *Physalospora psidii*

Symptoms

Affected twigs show wilting and death. Cracks and lesions are formed along the stem, arresting translocation of nutrients. Infected fruits turn dark brown to black and dries up resulting in die-back Symptoms. Fruit rotting takes place, blighting of leaves to enlargement.

Pathogen: *Physalospora psidii* Stev. & Pier. Perithecia is glabrous with a fleshy wall. Ascospores are hyaline, narrow, ellipsoid and one celled. Conidia are single celled, ovoid with a rough wall and measure 20 to 26 x 9 to 12 jam. On the stems and fruits pycnidia are formed in stroma.

Mode of spread and survival: The pathogen remains in the infected tissues beneath the bark and become active under favorable conditions.

Management

1. In severe infection, the disease can be prevented by the removal and destruction of the infected stem.
2. In mild infection, pruning of infected stem and branches is done and the cut-ends are painted with Bordeaux paste (1 part copper sulphate and 2 parts each of lime and linseed oil) or Chaubatia paste (copper carbonate - 800 g, red lead - 800 g and linseed oil - 1 litre).
3. Spraying the trees with copper oxychloride 0.2 per cent after pruning reduces canker incidence.
4. Anthracnose/Die-back/Fruit spot/Twig blight

Symptoms

The disease attacks all plant parts except roots. Severity of the disease may show die-back of main branches resulting in death of plants. The most characteristic Symptoms appear during the rainy season as small pin-head sized spots on the unripe fruits. They gradually enlarge to form sunken and circular, dark brown to black spots. The infected area of the unripe fruits become harder and corky. Acervuli are formed on fruit stalks.

Pathogen: *Gloeosporium psidii* Delacr. *(Perfect stage: Glomerella psidii* (Del.) Sheld.*)* Conidia are hyaline, aseptate, oval to elliptical or straight, cylindrical, obtuse apices or flattened at base. Conidiophore is cylindrical and tapers towards apex. It is hyaline and septate with single terminal phialide. Acervuli are dark brown to black.

Mode of spread and survival

The pathogen remains dormant for about three months in the young infected fruits. It becomes active and incites rot when the fruit begins to ripe. In moist weather, acervuli appear as black dots scattered throughout the dead parts of the twigs. From the twigs, the fungus penetrates the petioles and attacks the young leaves, which become distorted with dead areas at margins or tips. The conidia are spread by wind or rain.

Epidemiology

The cool season (Jan - Mar) and the hot, dry weather (Apr-Jun) prevent the spread of infection. In moist weather, acervuli are produced in abundance on dead twigs and pinkish spore masses are seen. Conidia initiate fresh infection. The temperature for disease development on fruits ranges from 30 to35°C.

Management

- Spraying the trees with Bordeaux mixture 1.0 per cent or copper oxychloride 0.2 per cent or Carbendazim 0.1% before the onset of monsoon reduces the disease incidence.
- Apple Guava (light red fleshed) is moderately resistant to anthracnose.

5. Red rust: *Cephaleuros virescens*

This disease is exceptionally severe in guava.

Symptoms

The alga produces specks to big patches on the leaves. They may be crowded or scattered. The pathogen extends between cuticle and epidermis and penetrates the epidermal cells. Fruit infection by alga is not common on fruits. Fruit lesions are usually smaller than leaf spots. They are dark green to brown or black in colour.

Mode of spread and survival

The disease is more common on closely planted mother plants. The zoospores cause the initial infection. High moist condition favours the development of fruiting bodies of the alga.

Primary source of inoculum: Dormant mycelia

Secondary source of inocu,um: Zoospores

Management: This algal disease is controlled by spraying with Bordeaux mixture 1.0 per cent or copper oxychloride 0.3 per cent.

6. DISEASE OF SAPOTA

1. Leaf spot : *Phaeopleospora indica*, *Pestalotiopsis versicolor*

Symptoms

***Phaeopleospora indica*:** Earlier circular spots which pinkish then gradually to brownish in colour and the centre of the spot sometimes whitish grey colour and number of spots is more on leaves.

***Pestalotiopsis versicolor*:** spots are circular and brownish and bigger. Later stages can see the black dots on centre of the spot. These black dots are the asexual fruiting body of the fungus (Acervulus). In advanced stages leads to defoliation.

Management: Carbendazim 0.1% and Companion (Combi product) includes carbendazim and mancozeb 12% and 72% to avoid resistant development in pathogen.

2. Flat limb: *Botyodiploidia theobromae*

Symptoms

In young stems instead of normal growth flattening takes place. On this flatted stem can see the small sized leaves with small petioles. This is a sporadic disease in plant 1 or 2 branch in whole plantation 1 or 2 plants are affected.

Management

Cut the affected stems and burn it while paste the cut portion with COC 0.3% to avoid dieback.

3. Sooty mould: *Capnodium versicolor*

Symptoms

Disease severity increases in increased population of leaf hoppers, aphids and other insects. Black superficial growth on entire surface of leaves, fruits and twigs. Fungus is not a parasite. It grows on the excreta and honey secretions of insects. Under dry spell such affected leaves curl & shrivel. During flowering time the appearance of the disease results in reduced fruit set. Sooty mass is a superficial growth of the fungus and it multiplies on insect secretions. Impact of this disease on host is photosynthesis activity and yield decreases.

Primary source of inoculum: Dormant mycelia

Secondary source of inoculum: Air borne conidia: Spread: Insects, Aphids, wind

Epidemiology: Temperature 28 -32°C, 85-90% RH, Warm Weather and susceptible host

Management

1. Spray wettable sulphur 0.2% along with insecticide Dimethoate 1.5g/lit.
2. Spray of 1% starch solution.

4. Red rust : *Cephaleuros versicolor*

The algal disease and it has been observed in India and elsewhere. It is one of the minor disease of importance. Reduction in photosynthetic activity and defoliation as a result of algal attack lower vitality of the host plant.

Symptoms

The disease is characterized by initial green coloured, circular patches with marginal serrations. The upper surfaces of the spot consist of numerous, unbranched filaments, which project through cuticle. As and when disease advances the organism turns red rusty spots on the leaves and young twig. Spores mature, fall off and leave cream to white velvety texture on the surface of leaf.

Epidemiology

The disease is more common on close plantation. The zoospores cause initial infection. High moist condition favours development of fruiting bodies of the algae.

Management: it is controlled by spraying with Bordeaux mixture 1% or Copper Oxychloride 0.3% or lime sulphur 0.2%.

7. DISEASES OF PAPAYA

1. Stem rot/foot rot: *Pythium aphanidermatum* (Eds.) Fitz, *Rhizoctonia solani* Kuhn.

Symptoms

Disease may lead to complete failure of crop when it appears in the early stages of growth. Water soaked patches appear on stem at ground level. These patches enlarge and girdle the base of stem. Diseased tissues turn dark brown or black & red. Terminal leaves turn yellow, drop & wilt. Fruits shriveled and drop prematurely. Disintegration of parenchymatous tissues at the base of stem takes place. The entire plant topples over the ground. The internal tissues dry up and give a honeycomb appearance. Rotting may spread above and below on the stem and down to the roots. The roots deteriorate and may be destroyed.

Disease cycle

The fungi survive in the form of oospore (Pythium) and sclerotia (Rhizoctonia) in the soil. This may serve as primary source of inoculum. The infected seedlings raised in infected soil carry the disease to the field. Secondary spread takes place through sporangia and zoospores.

Favorable condition

One week old seedlings are more susceptible than one year old trees. Stem rot caused by *P. aphanidermatum* is commonly noticed in 2 to 3 years old trees. The disease appears during rainy season and severity increase with the intensity of rainfall opt. temp. 36° C is favorable for disease development. *Rhizoctonia solani* is severe during dry & hot weather - 30° C - 35 ° C temperature.

Management

1. Seed treatment with captan or chlorothalonil @ 4 g/kg of seed
2. Seedlings should be raised on well-drained nursery area.
3. Diseased seedlings should be uprooted & destroyed.
4. Drenching the tree basin with Bordeaux mixture @ 1.0% or captan @ 0.2% or copper- oxycholoride @ 0.25 % or metalaxyl @ 0.1 % reduces the incidence of the disease.

2. Phytophthora Blight of Papaya: *Phytophthora palmivora*

Symptoms

Young fruits: Water-soaked lesions. Exude milky latex. Fruits may eventually mummify and fall.

Mature fruits: Fruit rot initially appears as small, circular, water-soaked lesions about (5–10 mm) in diameter. Large lesions, often forming first where the fruit contacts the stem of the plant are covered with whitish mycelium and masses of *Phytophthora* sporangia. Fruits can rot, turn soft, and fall prematurely.

Stems and foliage: The top portion of the fruit-bearing region of the stem is susceptible to infection during rainy periods. This can cause stem cankers to appear. The infected plant may become more susceptible to wind damage. Older portions of stems are susceptible when wet after extended rainfall, or after injury. As lesions enlarge, infected areas of the stems may weaken, causing stem damage or breaking. Foliage on affected stems may collapse.

Roots: Lateral roots of young plants (less than 3 months old) are most susceptible in poorly drained soils. Roots may become dark and rotten, causing stunting of plant growth and yellow, collapsed leaves. Severely infected plants may die. Plants with a heavy load of fruit may topple. Papaya plants with rotten roots are susceptible to drought stress.

Favourable condition and spread: The minimum temperature for growth of *P. palmivora* in culture is 52°F (11°C). The optimum temperature is 81.5–86°F (27.5–30°C), and the maximum growth temperature is near 95°F (35°C).

Propagules of this pathogen are dispersed principally by wind-blown rain, splashing rain, slugs, ants, knives, clippers, rodents, soil, or plant growth media.

Management

- Remove and destroy fallen fruits, especially those with disease symptoms.
- Spray pesticides, sometimes as often as weekly to biweekly to control Phytophthora diseases on papaya especially before and during rainy periods.
- Select a low-rainfall site for cultivation of papaya.
- Intercrop papaya with non-susceptible host plants.
- Do not grow papaya crops successively in the same field.
- Use non-infested soil or media for new transplant holes in fields with Phytophthora root rot (this is known as the "virgin soil" technique).
- Seedbeds in nurseries should be steamed or fumigated prior to planting.
- Control incipient rots (less than 24 hours old) of harvested fruit by dipping fruits in hot water held at 120°F (48°C) for 20 minutes.
- Avoid damage or injury to papaya stems during cultivation.
- Control African snails; they can vector the pathogen.
- Spray COC or Metalaxyl for curative.

3. Powdery Mildew: *Odium caricae*

Symptioms

The development of powdery mildew in papaya is promoted by high humidity (80-85%) and a temperature range of 24-26°C. The disease appears as on the foliage and pods. Infection is first apparent on the leaves as small slightly darkened areas, which later become white powdery spots. These spots enlarge and cover the entire leaf area. Severely infected leaves may become chlorotic and distorted before falling. Affected fruits are small in size and malformed.

Control : As soon as the disease Symptoms are observed dusting Sulphur (30 g/10 litres of water) or spraying hexaconazole 5 EC (10 ml/10 litres of water) at 15 days interval helps to control the disease.

4. Anthracnose: *Colletotrichum gloeosporioides*

Symptioms

The disease prominently appears on green immature fruits. The disease symptoms are in the form of brown to black depressed spots on the fruits. The initial symptoms are water-soaked, sunken spots on the fruit. The centres of these spots later turn black and then pink when the fungus produces spores. The flesh beneath the spots becomes soft and watery, which spreads to the entire fruit. Small, irregular-shaped water-soaked spots on leaves may also be seen. On the fruits, the symptoms appear only upon ripening and may not be apparent at the time of harvest. Brown sunken spots develop on the fruit surface, which later on enlarge to form water soaked lesions. The flesh beneath the affected portion becomes soft and begins to rot.

Control

- The affected fruits should be remove and destroyed. The fruits should be harvested as soon as they mature.
- Spaying with Copper Oxychloride (3 g/litre of water) or Carbendazim (1 g/ litre of water) or Thiophanate Methyl (1 g/litre of water) at 15 days interval effectively controls the disease.
- Fruits for exports should be subjected to hot water treatment or a fungicidal wax treatment.

5. Mosaic: Papaya Mosaic Virus (PaPMV), Carica Virus – 1

Symptoms

Disease may appear at any stage of crop growth but most serious on young plants. Typical mosaic Symptoms showing chlorosis with dark green blisters on leaves. Top young leaves of diseased plant are much reduced in size and show blister like patches of dark green tissue, alternating with yellowish green lamina and puckering. The leaf petiole is reduced in length and top leaves assume a upright position. On diseased fruits circular water soaked lesions with central solid spot appears. Fruits are deformed elongated and reduced in size and show mosaic patches. No reduction in the flow of latex.

Virus: Virus particles are filamentous 530 nm long. Thermal inactivation point 50-53° C, dilution end point 10^{-3} to 10^{-4}.

Mode of spread and survival: Not transmitted through seed of infected fruits. The virus is transmitted by sap, grafting and several aphids-*Aphis gossypii* & *Aphis medicaginis.*

Management

- Use healthy seedlings for planting.
- Rouging of infected plant & destroying them.
- Spraying of systemic insecticide for checking the spread of vectors by Acephate 75% SP @ 1 gm/Litre of water or Dimethoate 30% EC @ 1.7-2 ml/Litre of water or Fipronil 5% SC @ 1.5 ml/Litre of water

6. Leaf curl: Tobacco leaf curl virus (TLCV)

Symptoms

The disease is characterized by severe curling, crinkling and distortation of the leaves accompanied by vein clearing and reduction in leaf lamina. The leaf margins are rolled downward and inward in the form of inverted cap. Veins thickened and turn dark green. Leaves become leathery and brittle and petioles are twisted. Affected plants bear only a few flowers and fruits. The plant become stunted and leaves get defoliated

Mode of spread: The disease is transmitted through white fly *Bemisia tabaci* and not sap transmissible. **Host:** Tobacco, tomato, sun hemp, goose berry, chilli, hollyhock, zinnia and several other weeds.

Management

- Infected plants should be destroyed from the nursery
- In orchard , the infected plants roughed and destroyed
- Spray Imidacloprid 17.8% SL @ 0.2 to0.3 ml/Litre of water or Acephate 50% + Imidacloprid 1.8% SP @ 0.75-1 gm/Litre of water or Buprofezin 25% SC @ 1.5 ml/Litre of water or dimethoate 30% EC @ 1.7-2 ml/Litre of water for controlling vectors

7. Ring spot: Papya Ring Spot Virus (PRSV)

Symptoms

Disease is characterized by vein clearing and puckering. The margins and distal parts of young leaves roll downwards and inwards The plant get stunted. On the stem of young plants, mosaic or mottle symptoms with dark green spot and oily or water soaked streaks. Fruits are smaller, showing typical circular and concentric rings. Diseased fruits contain 40% lower sugar. Latex quality of diseased plants is poor. The virus particles are rod shaped, length 760-800 nm with 12 nm width.

Control

- Early detection of infected plants and prompt removal can check the spread of the disease.
- Spraying of systemic insecticide for checking the spread of vectors by Acephate 75% SP @ 1 gm/Litre of water or Dimethoate 30% EC @ 1.7-2 ml/Litre of water or Fipronil 5% SC @ 1.5 ml/Litre of water.
- Spraying of neem oil 0.1%

8. DISEASES OF POMEGRANATE

1. Cercospora leaf spot

Symptoms

Light brown zonate spots appear on the leaves and fruits. Black and elliptic spots appear on the twigs. The affected areas in the twigs become flattened and depressed with raised edges. Such infected twigs dry up. In severe cases the whole plant dies.

Causal organism: *Cercospora punicae* (P.Henn). Conidiophores are olivaceous brown, short fasciculate, sparingly septate. Conidia are hyaline to pale olivaceous cylindrical to sub clavate and septate.

Mode of survival and Spread: The pathogen survives in affected plant parts as dormant mycelia and spreads through airborne conidia produced in acervullus.

Epidemiology: The disease is severe during August to November. When there is high humidity and the temperature between 20 and 27°C.

Management

Cultural practice

- Clean cultivation, i.e. sanitation, includes removal of weeds.
- On fallen leaves or affected plant parts, spray nitrogen solution or bleaching powder to enhance degradation.
- Prune all affected branches then burn affected branches and pruned material and pruned area should be smeared with Bordeaux paste or COC paste.

Chemical: Spray thiophanate methyl 0.1% or mancozeb 0.2% or cardendazim 0.1%.

2. Bacterial blight

Symptoms

Small irregular, water soaked spots appear on the leaves. Spots vary from two to five mm in diameter with necrotic centre of pin-head size. Spots are translucent, later turn light brown to dark brown and are surrounded by prominent water-soaked margins. Spots coalesce to form large patches. Severely infected leaves fall off. The bacterium attacks stems, branches and fruits also. On the stem, the disease starts as brown to black spots around the nodes it leads to girdling and cracking of nodes. Finally the branches break down. Brown to black spots on fruits is raised and oily in appearance.

Causal organism: *Xanthomonas axonopodis* pv.*punicae* [= *Xanthomonas campestris* pv.*punicae*]. It is Gram-negative rod, motile with single polar flagellum. It is non acid fast and aerobic.

Mode of spread and survival

The bacterium survives on the tree. The pathogen survives for 120 days on the fallen leaves during the season. The primary infection is through infected cuttings. The disease spreads through wind splashed rains.

Epidemiology: High temperature and low humidity favor the disease. Temperature of 30 to 34°C, relative humidity of 80 to 85% is favorable for multiplication of pathogen.

Management

- Clean cultivation and strict sanitation in the orchard help to reduce the disease incidence.
- Collect and burn the fallen leaves.
- Spraying of 1 % urea solution to fallen leaves enhances the degradation.
- Bleaching on to the fallen leaves reduces the inoculum.
- Spraying the Bordeaux mixture 1.0% controls the disease.
- Spray 0.05% streptocycline to control the disease.
- Also can use copper oxy chloride spray at 0.3% concentration.
- Pruning at correct stage would reduce the disease(Bahar pruning).
- Ganesh as moderately resistant variety for bacterial blight disease.

3. Leaf spots: *Colletotrichum gloeosporioides*

Symptoms

The disease appears as small, regular to regular to irregular dull violet or black spots on the leaves. These spots are surrounded by yellow margins. The infected leaves turn yellow and drop off.

Mode of survival and Spread: The pathogen survives in affected plant parts as dormant mycelia. Spreads through airborne conidia produced in acervullus. Mode of entry through stomata.

Epidemiology: The disease is severe during August to November. When there is high humidity and the temperature between 20 and 27°C.

Management

- Clean cultivation.
- On fallen leaves or affected plant parts, spray nitrogen solution or bleaching powder to enhance degradation.

- Prune all affected branches then burn affected branches and pruned material and pruned area should be smeared with Bordeaux paste or COC paste.
- Spray thiophanate methyl 0.1% or mancozeb 0.25% or cardendazim 0.1%

4. Fusarium wilt: *Fusarium fusarioides* (Farg & Cif)

Symptoms

The disease appears as minute specks towards the leaf margin. The spots are brown, circular to irregular in shape. Later the spots coalesce and form big dark brown necrotic blotch.

Mode of survival and Spread: The pathogen survives in affected plant parts and spreads through airborne conidia.

Epidemiology: The disease is severe during August to November. When there is high humidity and the temperature between 20 and 27°C.

Management

- Clean cultivation, i.e. sanitation, includes removal of weeds. On fallen leaves or affected plant parts, spray nitrogen solution or bleaching powder to enhance degradation.
- Prune all affected branches then burn affected branches followed by smeared with Bordeaux paste or COC paste.
- Spray thiophonate methyl 0.1% or mancozeb 0.2% or cardendazim 0.1%.

9. DISEASES OF BER

1. Powdery mildew-*Oidium erysiphoides* f.sp. *zizyphi*

Symptoms

The developing young leaves show a white powdery mass causing them to shrink and defoliate. Small, white powdery growth appear on the young fruits which later enlarge and coalesce and final turn brown to dark brown. In severe cases, the whole fruit surface gets covered with the powdery mass. Affected young fruits drop off prematurely or become corky, cracked, misshapen and underdeveloped. Matured fruits turn rusty. Sometimes the whole crop is rendered unmarketable .

Etiology: Conidiophores are white upright. Conidia are cylindrical, single celled, hyaline, barrel shaped and 22.5 to 37.8 × 16.8 to 21.0µm.

Perpetuation: spread through wind born conidia

Epidemiology: Warm high humid conditions with relative humidity more than 90% favors diseases development.

Management

- Spraying of hexaconazole 0.1 per cent or wettable sulphur 0.25 per cent should be done during first and third weeks of November or when the fruit attains pea size. Triton-AE or Teepol or Sandovit may be added for adhesion.

2. Alternaria Leaf spot - *Alternaria chartarum*

Symptoms

The disease is characterised by the formation of small irregular' brown spot on the upper surface of the leaves. On the lower surface dark brown to black spots are formed. The spots coalesce to form big patches. The diseased leaves later drop. Plant debris serve as potential source of primary infection.

Perpetuation: Plant debris serve as potential source of primary infection. Secondary spread is through the wind born conidia.

Epidemiology: The disease development is favored at 20 to 30 ^{0}C with an optimum of 25^{0}C. High humidity and frequent rainfall seem to be more important than temperature for disease development.

Management

- Spray 0.1 per cent propiconazole or 0.1 per cent azoxystrobin.
- Resistance variety namely Bahadurgarhia, Goal Gurgaon, and Poular Gola and ZG3 may be grown.

3. Rust - *Phakopsora zizyphi-vulgaris*

Symptom

On the lower surface of the leaves small, irregular, reddish brown uredopustules appear. Later they cover the whole area of the leaves. The infected leaves dry and defoliate.

Etiology: Natural occurrence of teleutospores and formation of basidia and basidiospores after germinations of teletuospores have been recorded and this cause infection to plants.

Management: It is an autoecious rust. Spraying with Mancozeb 0.2 per cent or Zineb 0.2 per cent or Wettable sulphur 0.2 per cent.

Soft rot:*Phomopsis natsume*

The disease appears as a light russet vinaceous coloured, irregular spot on the fruits. It increases in size and make the whole fruit into pulpy, brown to black in colour with soft and loose outer skin.

Management: The disease can be controlled by spraying with Carbendazim 0.05 per cent.

10. DISEASES OF STRAWBERRY

1. Powdery mildew

Symptoms

Whitish powdery growth on upper surface of the leaf even young stem defoliation fruit cracking take place also fruit size reduction reduced fruit yield. Occasionally a powdery or surface mildew cause s some damage to plants.

Casual Organism: *Sphaerotheca fragariae*

Etiology

Mycelium is white, septate, ectophytic and sends globose haustoria into the epidermal cells of the host. Conidiophores are short and erect. Conidia are one celled, oblong, and minutely verrucose with many large fat globules. Cleistothecia are formed towards the end of the season on the leaves, petals, stems and thorns. Cleistothecia are with simple myceloid appendages. Each ascus contains 8 ascospores.

Mode of spread and survival

The fungus over winters mycelium in dormant buds and shoots which are not entirely killed.Either conidia or ascospores serve as primary inoculum. Secondary spread is through wind borne conidia. (Cleistothecia of the fungus)

Epidemiology : Infection occurs when the air is saturated with moisture and the temperature is about 200c. Optimum conidial germination occurs at 97 to 9per cent relative humidity and at temperature ranging between 17 to 24°c. Spore production is maximum at 24 to 28°c.

Disease cycle

Affected leaves, buds and twigs having clestothecia, in favourable conditions it produce ascus (sexual fruiting body) in that ascospores are present. Ascus liberates ascospores and they flight, landing on to the host surface and cause infection leeds to powdery mildew. White powdery growth comprising of oidea. Oidea releases barrel shaped conidia and cause infection and continues asexually life cycle. Adverse climatic conditions the fungus switched on to sexual stage by production of Antheridium and Ascagonium. Gametangial contact type of reproductiononce Ascogonium and Antheridium come in contact together, plasmogamy, karyogamy, meiosis followed bymitosis and ascospore formation take place.

Management

- The diseased and fallen leaves should be collected and burnt.
- Four sprayings at 1day interval with wettable sulphur 0.3 per cent or hexaconazole 0.1 per cent or carbendazim 0.1 per cent controls the disease effectively.
- Spraying with Phalton 0.3 per cent + Carbendzim 0.1% also controls the disease. Spraying with triademefon (bayleton) 0.1 per cent at 3days interval controls the disease.
- Some of the resistant varieties are Aawliver, Abhisarika, Adolf morstman, African star, Barbara etc.
- Excess fertilization especially with nitrogenous fertilizers and crowding of plants should be avoided.

2. Leaf Spot

Casual Organism: ***Mycosphaerella fragariae***

Symptoms

Leaf spot is most frequently evident on the blades of the leaflets, but may appear on the petioles, fruit, and fruit stems. The lesions may be seen first on the upper surface as small, deep-purple, somewhat indefinite areas. As the spot enlarges, the central area becomes brown, but soon turns to a definite white spot in older leaves or to a light brown in young tender leaves. An indefinite dark purple zone surrounds the central light area, giving the whole a birds-eyes effect. When the infections are bunched on the leaf, the purplish area may become confluent and extend around a number of the white spots, and if the infection is near the edge of the leaflet, the purpling often extends to the border. One the undersurface the symptoms are much the same as on the upper, but the coloring is less intense. Here the prominent veins which are touched by any of the spots take on a reddish-purple color which extends some distance beyond the infected spots.

Management

- Curing of planting material.
- Spray Copper Oxychloride 0.3% or Carbendazim 0.1%.

3. Strawberry Leaf Scorch

Causal Organism : ***Leptothyrium fragariae***

Symptoms

Leaf scorch lesions may appear not only on the leaf blades but also on the petioles, fruit pedicels, and on the sepals of the calyx. In a very early stage leaf scorch lesions resemble those produced by the leaf spot organism in that small dark purple spots appear scattered over the upper surface of the leaflets. It is not difficult to distinguish the two after the spots have developed. In the mature condition the leaf scorch spots are large and irregular in outline and never show the white central area characteristic of the leaf spot disease. On the contrary, the black fruiting bodies which develop in the central area give the leaf scorch disease a tar spot appearance.

Disease Cycle

Strawberry leaves often survive the winter, and diseased leaves may be found in the early spring bearing both the perfect and imperfect stages of the leaf scorch organism. Ascospores are more important than the conidia in primary infection, since they are discharged in great numbers during the early spring months when the new leaves are developing. Under suitable moisture conditions the ascospores germinate within 24 hours and infection takes place by direct penetration of the epidermal cells, in contrast to the stomatal infection of *Mycosphaerella fragariae*.

Management

- Use Healthy planting material.
- Avoid creation of un necessary wounds.
- Application of fertilizer should be delayed at least 1days.
- spraying with Copper oxychloride 0.3 per cent or Chlorothalonil 0.2 %.
- Use of resistant varieties.

4. Leaf Blight

Casual organism: *Dendrophoma obscurans*

Symptoms

The disease is most conspicuous on the leaves, although at times it appears on the calyx. Usually the spots on the single leaflet are limited to one to five or six. When first observed, the young spots are uniformly reddish purple and almost circular in outline. If they are near one of the main veins, the spots are elliptical. Later three zones may be normal green of the leaf, a light brown zone about 5 mm width, and, finally, a dark brown central area 2 to 3 mm in diameter. The white central area characteristic of leaf spot is never present. If the spots occur on a prominent vein, and especially if on the midrib, the typical V-shaped lesion is formed, with the purpling of the tissue extending fanlike to the border of the leaflet.

Management

Spray 0.1% carbendazim

5. Cortical Root Rot (Black Root) of Strawberry

Symptoms

A plantation affected by root rot presents an uneven patchy appearance due to dwarfing of the diseased plants and to gaps caused by the complete death of the severely affected plants. It is generally agreed that certain definite concerned. Under certain environmental conditions some of these organisms are considered much more virulent than others.

Management

- Biological control: *Trichoderma viride* @ 0.6%.
- Chemical: Carbendazim @ 0.1% spray.

6. Crown rot of strawberry

Symptoms : The fungus *Pestalotiopsis* sp. caused crown rot with typical symptoms are drying starts from the edges of the leaves and lesions appear first on leaves and spreads down to crown region. The severely infected plants have completely dried leaves and flowers then turned dark in color, stems and roots are black with disease spreading over the all runners.

Casual Organism: *Pestalotiopsis clavispora* (G. F. Atk.) Steyaert.

Management

- Use Healthy planting material.
- Avoid creation of unnecessary wounds.
- Use recommended dose of fertilizer.
- Spraying with tebuconazole 0.1 per cent and drench *Trichoderma harzianum* 0.6 per cent.
- Use of resistant varieties.

References

Chadha, K.L. 2002. Hand Book of Horticulture. ICAR, New Delhi.

https://agrimoon.com/disease-of-horticultural-crops-their-management-pdf-books/

Pathak, V.N. 1980. Diseases of Fruit Crops. Oxford IBH Publishing Co. Pvt. Ltd., New Delhi.

Ranga Swamy, G. 1988. Diseases of Crop Plants in India. Prentice Hall of India Pvt. Ltd., New Delhi.

Singh, R.S. 2000. Disease of Fruit Crops. Oxford & IBH Publishing Co. Pvt. Ltd. New Delhi.

Thind, T.S. 2001. Disease of Fruits and Vegetable and Their Management. Kalyani publishers, Ludhiana.

Tiwari, R.K.S., Singh, A.K., Nrmalkar, V.K. and Gupta, C.R. (Eds.), 2014. Disease of Horticultural Crops and their Management. Available at: www. Igau.edu.in

Verma, L.R. and Sharma, R.C. 1999. Diseases of Horticultural Crops. Indus Publishing, New Delhi.